USING COUNTER-EXAMPLES IN CALCULUS

USING COUNTER-EXAMPLES IN CALCULUS

John Mason

Open University in Milton Keynes, UK

Sergiy Klymchuk

Auckland University of Technology, New Zealand

ICP Imperial College Press

Published by

Imperial College Press
57 Shelton Street
Covent Garden
London WC2H 9HE

Distributed by

World Scientific Publishing Co. Pte. Ltd.
5 Toh Tuck Link, Singapore 596224
USA office: 27 Warren Street, Suite 401-402, Hackensack, NJ 07601
UK office: 57 Shelton Street, Covent Garden, London WC2H 9HE

British Library Cataloguing-in-Publication Data
A catalogue record for this book is available from the British Library.

USING COUNTER-EXAMPLES IN CALCULUS

ISBN-13 978-1-84816-359-1
ISBN-10 1-84816-359-2
ISBN-13 978-1-84816-360-7 (pbk)
ISBN-10 1-84816-360-6 (pbk)

Typeset by Stallion Press
Email: enquiries@stallionpress.com

Printed in Singapore.

PREFACE

This book is intended as supplementary material for enhancing teaching/learning of first-year university courses in calculus. It can also be used in upper secondary school. Chapters 1 and 2 explore pedagogical aspects of working with counter-examples. Chapter 3 consists of carefully constructed incorrect mathematical statements for students to puzzle over and to try to create their own counter-examples in order to disprove them. Chapter 4 provides well illustrated solutions to the practical exercises of Chapter 3 and offers new challenges in the form of questions for discussion. Many of the statements have been chosen because they reflect common misconceptions that students pick up or construct. Some of the false statements are converses of famous theorems or true facts, some are created by omitting or changing conditions of the theorems, some are incorrect definitions and most appear at first glance to be correct. The following major topics from a typical course on Calculus of a single variable are considered: Functions, Limits, Continuity, Differential Calculus and Integral Calculus.

The book offers much more than a collection of prompts to stimulate students to construct counter-examples. It promotes an approach to teaching which sees stimulating learners to use their own natural powers as both motivating and empowering. Even where students are taking calculus for purely pragmatic purposes, we believe that their interest and involvement can be enhanced by engaging them in the activity of example construction. If they do not appreciate the scope and range of the mathematical objects they are supposed to be learning about, then they are ill prepared to use the techniques in other situations. The book offers advice for students on how to go about constructing counter-examples, and advice on what to do with examples when they are provided, encountered, or discovered.

The thrust of the book follows proposals of Watson and Mason (2005) inspired by authors such as George Polya, Paul Halmos and others. They advocate getting learners to construct not just one but classes of examples for themselves in order both to extend and enrich their own example spaces, and to develop a full appreciation of concepts, definitions and techniques that they are taught. The book therefore offers the reader stimulation to work on their own mathematics together with pedagogic advice on how to make best possible use of examples and counter-examples. Each statement in Chapter 3 is associated with one or more counter-examples in Chapter 4, and many also have indications of how these counter-examples can be generalised to a broader class. At a deeper level, the book provides an array of strategies for exposing learners to the role of counter-examples and to their construction. A by-product is a collection of useful functions for learners to use as test items when they meet new assertions.

This book follows in the footsteps of books with similar titles, but it is at once more elementary in the mathematics it presents, and more sophisticated in its pedagogical constructs and ways of working. For example, the well-known book on counter-examples in Calculus: "Counter-examples in Analysis" by B. R. Gelbaum and J. M. H. Olmsted (Holden-Day, Inc., San Francisco, 1964) is an excellent resource for teaching/learning of Calculus at an advanced level but it is well beyond the scope of a basic first-year university Calculus course. Other similar titles are at upper undergraduate and graduate level. None of them say anything about what a student might do with the counter-examples provided so as to develop their mathematical thinking. There is no overlap between this book and "Counter-examples in Analysis" in either statements or examples. This book fills a gap in the teaching of first-year university Elementary or Introductory Calculus, providing both tasks and pedagogy to enhance and enrich student's appreciation of topics. It can be useful for:

— upper secondary school teachers and university lecturers as a teaching resource;
— upper secondary school and first year university students as a learning resource;
— upper secondary school teachers for their professional development in both mathematics and mathematics education.

Many of the examples used here as counter-examples actually arise in engineering or scientific situations. They are neither weird nor outlandish. For mathematics students the examples used contribute a valuable addition to their example-spaces, and the strategies suggested add to their toolkit of ways of working on mathematics. For teachers and lecturers of engineering, science and mathematics students, they provide a valuable resource on which to draw while teaching.

CONTENTS

INTRODUCTION

Many students are quite happy to apply techniques to situations in which the necessary conditions for the appropriate use of those techniques fail to hold. A reasonable conjecture is that students focus on numerical or computational aspects of techniques and overlook verbal conditions. This is especially the case where the conditions appear highly technical, or are identified with "what is always the case anyway".

Many students are unused to thinking about problems before diving in and doing whatever comes to mind first. They make a wide range of slips and errors without even noticing them, often because they are not in the habit of thinking first, nor of checking their ideas and conjectures on examples and using this experience to modify and correct their intuitions.

Many students have limited example spaces available to them for seeking inspiration, for testing out conjectures, or for use in applications. Lacking familiarity with all but a very few examples, they have little appreciation of the scope of mathematical objects to which the theorems and techniques apply.

This book provides an antidote to these phenomenona. Focusing student attention directly on necessary conditions, and getting them to construct or locate counter-examples when the conditions are adjusted, serves to illuminate the need for those conditions. Naturally enough, being directed to seek counter-examples has been found to focus learner attention on relevant features of examples (Wilson, 1986).

Helping learners appreciate that single counter-examples are representative of infinite classes of examples means that learners are less likely to "monster-bar" in the sense of Lakatos (1976): ignore the example as being isolated and irrelevant to their real concerns. As a result, learners are more likely to appreciate that these examples are neither isolated nor pathological, but part of the extensive domain of generality to which the theorems and techniques apply. By using the strategies suggested, teachers can enculturate learners into checking that necessary conditions are satisfied before using the technique or applying the theorem, as well as alerting them to check their work for slips and errors.

There is an ongoing vibrant argument between those who advocate exposing students to "pathological examples" and those who advocate keeping things as simple as possible for their students. The issue is a complex one and is taken up in Chapter 2. While acknowledging some of the reasons for the other position, our aim is to provide in the intervening chapters a convincing pedagogical case for extending students' example spaces. We hope that in this way our arguments in favour of the use of counter-examples will be found more convincing.

Examples and Counter-Examples

Examples are used to illustrate; counter-examples are used to disprove. But every counter-example can be seen as an example of a slightly different statement, and every example can be seen as a counter-example to at least one associated conjecture. Thus a mathematical object, whether number, function, or the result of applying operations to these, is in itself neither an example nor a counter-example. It all depends on context, setting, and intention. It can be an example of a range of different statements and a counter-example to a range of different conjectures.

There is a long and ancient history (some 4000 years!) of teachers using examples in order to help students appreciate and understand concepts being defined, theorems being proved, and techniques being demonstrated. However there is an equally long history of students using worked examples as mere templates for doing exercises, and for treating examples as single items rather than representatives of classes of objects. For example it is common for students to be shown the function $|x|$ as an example of a function which is continuous, and differentiable everywhere except at a single point. As Des McHale (1980) pointed out, students often treat this as a singularity and "monster bar" it. Since it is not a familiar object they can dismiss it as being irrelevant. However it is neither singular nor irrelevant, since it is intended to illustrate one of several things that can go wrong at a single point in the relation between continuity and differentiability.

Zazlavsky and Ron (1998) found in their study that "students' understanding of the role of counter-examples is influenced by their overall experiences with examples. The status of a counter-example is so powerful compared to the status of other examples". But those students have to have made a commitment to seeking what is true, what is the case.

Peled and Zazlavsky (1997) make useful distinctions between counter-examples, according to the degree of generality they display. They use the adjective *specific* for counter-examples which contradict the claim, but which give no indication as to how one might construct similar or related counter-examples; *semi-general* if it provides some idea of how one might generate similar or related counter-examples, but does not tell "the whole story" or does not generate the whole space of counter-examples; *general* if it provides insight as to why a conjecture is false and suggests a way to generate an entire counter-example space. These latter examples have the property of being relatively *transparent* counter-examples. But transparency has to do not with the example itself, nor even with how it is presented, but with how the student perceives the example (for example as singular or as generative of a class). If the presentation does little or nothing to draw attention to the aspects which make the example exemplary or which make it a counter-example, students are likely to miss the point and dismiss the example, as MacHale suggested.

Sometimes a conjecture has a single or a restricted class of counter-examples. In this case it makes sense to adjust the conjecture in order to rule out the examples. "Monster barring" can be a constructive activity in the struggle to find a succinct but elegant and effective statement of a definition so as to make desirable theorems easy to prove. For example, the definition of a fraction as the ratio of two integers has to include a clause which excludes division by zero; the function x^y has to exclude values of $x \leq 0$. Conversely, theorems are most powerful when they include as wide a range of objects as possible. For example, a theorem about the area or perimeter of a trapezium already includes the special case of a triangle. Seeing a triangle as a special case of a trapezium is an important perception to develop, just as at other times in school it is important to see both fractions (well, the value of fractions) and decimals as numbers, and to see squares as rectangles and rectangles as parallelograms.

If students are to appreciate the range and power of the techniques they are shown, then it is vital that they appreciate the possible objects which they admit by accepting definitions as stated. They also benefit from appreciating why definitions are framed the way they are, including the sorts of features which are desirable and the sorts which are to be excluded in order to prove theorems. As Lakatos suggested, the history of mathematics is replete with adjustments to definitions. A classic example of this is the definition of function. For Leonard Euler any relationship between two quantities x and y which could be expressed as a formula, constituted a function. Thus both $y = 2x+3$ and $y^2x = 2x+3$ expressed functions. Then there is the issue of glued functions such as

$$y = \begin{cases} x & \text{if } x \geq 0 \\ -x & \text{if } x < 0 \end{cases}$$

which can also be expressed as a formula using the absolute value, $y = |x|$. Many students prefer their functions to be specified by a single formula, and historically they are in good company! In order to become comfortable with functions formed by gluing formulae together at points, students can be asked to construct such functions so as to meet specified constraints. This strategy is developed in the next chapter.

Dahlberg and Housman (1997) also noted that their undergraduate subjects had trouble with the underlying concepts, e.g., function and root, making it hard to generate examples and non-examples of a made-up family of "fine functions" defined as functions which have a root at each integer. One student identified "root" with "continuity", three others initially thought the graph of the zero function was a point, and one did not believe the zero function was periodic. In addition, most students' initially thought in terms of functions which were non-constant polynomials or continuous (Selden and Selden, 1998).

Example Spaces

Most people, when they encounter a technical term, immediately look for a familiar example on which to test out what is being said. This is highly mathematical activity, but it is severely hampered if there are only a few examples to hand, or if the examples available do not display all relevant features. Even when you are fairly confident with specific concepts, theorems and techniques under normal conditions, when the weather turns rough and unfamiliar aspects are highlighted, it is really useful to be able to turn to familiar confidence inspiring examples in order to get a sense of what is happening. If the only examples to hand are relatively trivial, then they may not be sufficiently sophisticated to illustrate the difficulties. One of the reasons why students often do not pay attention to the conditions of a theorem or a technique is that the only examples with which they have any familiarity "naturally" fulfil all the conditions. If they have never encountered more extreme examples of the concepts involved, they may never appreciate why those conditions are required, in which case, the whole enterprise is likely to remain mysterious!

Watson and Mason (2005) found it helpful to think in terms of situations triggering access to a space of examples. Certain examples come to mind immediately. If these are individual and specific then the example space is rudimentary. If, instead, individual examples trigger awareness of ways in which those examples can be modified to produce whole classes of examples, the space is more richly interconnected. Ference Marton and colleagues (Marton and Booth, 1997; Runesson, 2005) speak of *dimensions of variation* to refer to aspects of examples which can be varied and still they remain examples. Watson and Mason observed that a single example gives access to whole classes of similar or related examples when the person is aware of the possibility of varying certain features. They also noted that lecturers and students often have different *dimensions of possible variation* in mind when pondering an example. Furthermore, even when the same dimensions are being considered, lecturers and students may differ in the *range of permissible change* of each dimension which they are allowing. For example, the lecturer may be thinking "any real number" while students are subconsciously restricting attention to integers, unaware even that they are making restrictive assumptions.

Example spaces are largely idiosyncratic and situation dependent. Over time the differences begin to iron out so that an expert is likely to develop strongly habituated examples that come to mind whenever a technical term is used. It is probably the case that different experts share certain examples in common, and these become the canonical examples offered in textbooks and in lectures. To be useful, it is vital that canonical examples are not allowed to be isolated, but rather provide access to broader classes through awareness of dimensions of possible variation and corresponding ranges of change (Goldenberg and Mason, 2008).

Edwina Rissland (née Michener, 1978) distinguished several uses for examples:

Start-up examples used to initiate study of a topic;

Reference examples used canonically for testing conjectures and illustrating techniques;

Model examples used to illustrate aspects and features, for example, $x(x^2+1)$ and $x(x^2-1)$ for cubics, $|x|$ for non-differentiability at a point, and $\sin(1/x)$ for discontinuity at a point and $(x^2-a)/(x-a)$ $(x \neq a)$ for a continuous function whose domain can be extended continuously to the whole of R.

Counter-examples used to demonstrate why conditions are required to be satisfied in theorems and techniques;

There is considerable slippage between some of these categories. In this book we concentrate on counter-examples because they are under used in teaching, thereby not only impoverishing students' experience and appreciation of the calculus but also making it difficult for them to appreciate what mathematics itself is about, and restricting their perception of the range of mathematical objects.

Mathematics as Constructive Activity

When mathematics is presented as a series of technical terms, theorems and techniques, students develop the impression that everything is worked out and that there is no room for creativity in mathematics. Yet every mathematical problem can be viewed as an opportunity to construct a mathematical object which resolves the problem, whether it be a set of numbers, a set of functions, or something even more sophisticated. Seen in this light, techniques are devices for making the construction, rather than handles to turn which pop out required answers. This view of mathematics as creative and constructive is even more important for those who are "only" going to use it in their own subjects.

For students studying calculus as a tool for use in their subject of interest, the situation is compounded because they tend to focus on the techniques in order to pass the assessment, without realising that an important feature of tools and techniques is knowing when to use them. Recognising routine situations is an easy matter. What is really useful is recognising their relevance in novel situations. To use the calculus effectively in constructing solutions to problems of interest, the mathematical techniques have to be used constructively and creatively. Where students have come to see mathematics as a creative and constructive endeavour, they are in a much more powerful position to use their knowledge creatively and effectively, either by exploring for themselves or seeking the assistance of an expert.

Students who have engaged in constructing mathematical objects meeting certain constraints, who have adopted a constructive view of mathematical thinking and problem solving will have much richer example spaces than colleagues who see mathematics as a collection of procedures, and are in a much better position to make effective use of mathematics in their own subject.

Students who have worked on the statements of theorems to discover what is *sharp* about the theorem, that is, why conditions are included, are more likely to check conditions before applying theorems and using techniques.

Reasons for Encouraging Students to Work with Counter-examples

We encourage students to work with counter-examples with the following purposes:

1. For deeper conceptual understanding

Many students have become used to concentrating on techniques, manipulations and familiar procedures, often under the direction of their teachers. They do not pay much attention to the concepts, conditions of the theorems, properties of the functions, or to reasoning and justification.

"When students come to apply a theorem or technique, they often fail to check that the conditions for applying it are satisfied. We conjecture that this is usually because they simply do not think of it, and this is because they are not fluent in using appropriate terms, notations, properties, or do not recognise the role of such conditions" (Mason and Watson, 2001). Paying attention to the conditions of theorems helps engineering students develop the good habit of considering the extreme conditions new devices will be subjected to. Aircraft are designed to fly in storms and turbulence, not just in perfect weather! The ability to pay attention to the conditions of a sale offer is essential in everyday life. We all know the importance of reading the fine print on advertisements "special conditions apply". A recent case study done in New Zealand (Klymchuk, 2005) indicated that the usage of counter-examples in teaching could significantly contribute to improving the students' performance on test questions that required conceptual understanding.

2. To reduce or eliminate misconceptions

Over recent years, partly due to extensive use of modern technology, the proof component of the traditional approach in teaching Calculus (definition-theorem-proof-example-application) has almost disappeared. Students are used to relying on technology and sometimes lack logical thinking and conceptual understanding. Sometimes Calculus courses are taught in such a way that special cases are avoided and students are exposed only to "nice" functions and "good" examples, especially at school level. This approach can create many misconceptions that can be explained by Tall's generic extension principle: "If an individual works in a restricted context in which all the examples considered have a certain property, then, in the absence of counter-examples, the mind assumes the known properties to be implicit in other contexts." (Tall, 1991).

In Chapter 3 of this book many false statements, given to disprove by counter-examples, are related to students' common misconceptions. There is a difference between students' misconceptions in basic algebra and in Calculus. There are no textbooks where "properties" like $\sqrt{a+b} = \sqrt{a}+\sqrt{b}$ can be found, and nobody teaches such "rules" either. Some introductory Calculus textbooks on the other hand, especially those at school level, contain incorrect statements. For example: "If the graph of a function is a continuous and smooth curve (no sharp corners) on (a, b) then the function is differentiable on (a, b)", and "a tangent line to a curve is a line that just touches the curve at one point and does not cross it there". Some students actually *learn* Calculus this way. Practice in creating counter-examples can help students reduce or eliminate such misconceptions before they become second nature.

3. To advance mathematical thinking

Creating examples and counter-examples is neither algorithmic nor procedural. It may require advanced mathematical thinking which is rarely experienced at school. "Coming up with examples requires different cognitive skills from carrying out algorithms since it is necessary to switch perspective and look at mathematical objects in terms of their properties. At first, to be asked for an example can be disconcerting, when students have not been exposed to example construction in action, and so have no pre-learned algorithms to show the "correct way" (Selden and Selden, 1998). Practice in constructing their own examples and counter-examples can help students enhance their creativity and advance their mathematical thinking.

4. To enhance generic critical thinking skills

Creating counter-examples to wrong statements has a big advantage over constructing examples of functions satisfying certain conditions, because counter-examples deal with disproving, justification, argumentation, reasoning and critical thinking, which are the essence of mathematical thinking. These skills will benefit students not only in their university study but also in other areas of life.

5. To expand the "example space"

After creating or being exposed to many functions with interesting properties students will expand their "example space", allowing them to better communicate their ideas in mathematics and in practical applications. While creating counter-examples students learn a lot about the behaviour of functions and can later apply their knowledge to solving real life problems.

For example, the counter-examples to Statement 2.2 and Statement 4.32 from Chapter 4 are the functions

$$f(x) = \frac{\sin x}{x} \quad \text{and} \quad f(x) = \begin{cases} x^2 \left|\cos \dfrac{\pi}{x}\right|, & \text{if } x \neq 0 \\ 0, & \text{if } x = 0 \end{cases}$$

respectively, which are used for modelling vibration processes in mechanical engineering; the counter-example to Statement 5.12 from Chapter 4 is the Fresnel function

$$F(x) = \int_0^x \sin t^2 dt$$

which apart from being important in optics has recently been applied to motorway design.

As Henry Pollak from Bell Laboratories, USA pointed out "society provides time for mathematics to be taught in schools, colleges and universities not because mathematics is beautiful, which it is, or because it provides a great training for the mind, but because it is so useful".

6. To make learning more active and creative

Experience of colleagues and our own teaching experience shows that the usage of counter-examples as a pedagogical strategy in lectures and assignments can create a discovery learning environment and make learning more active. A recent international study involving more than 600 students from 10 universities in different countries (Gruenwald and Klymchuk, 2003) showed that the vast majority of the participating students (92%) found the use of counter-examples to be very effective. They reported it helped them to understand concepts better, prevent mistakes, develop logical and critical thinking, and that they were more actively involved in lectures. Many commented that creating a variety of counter-examples enhanced their critical thinking skills in general, skills useful in other areas of life that have nothing to do with mathematics.

There are different ways of using counter-examples in teaching apart from disproving false statements:

- giving the students a mixture of correct and incorrect statements
- making a deliberate mistake in the lecture
- asking the students to spot an error on a certain page of their textbook
- giving the students bonus marks towards their final grade for providing excellent counter-examples to hard questions during the lecture
- revealing student confusions to inform future teaching
- as a form or aspect of assessment.

Chapter 1

WORKING WITH COUNTER-EXAMPLES

We take the view that technical terms and theorem statements trigger access to an example-space: A collection of illustrative examples and useful counter-examples associated with the technical term and-or theorem. The richer that space, the richer and more sophisticated your appreciation and understanding of the technical term. Example spaces accumulate and develop over time. They do not suddenly appear as the result of reading or attending a lecture. They need to be worked on and with.

A useful way of thinking about example-spaces is in terms of a pantry (Watson & Mason, 2004). Towards the front there are familiar examples used frequently. When an unusual object is encountered it may be stored away further back on the shelf in case it proves useful in the future. The important feature of an example-space is that the more familiar examples provide access to the less familiar objects stored further back on the shelf, so it is vital that examples are connected and related to each other when they are encountered. That is why it is important to do something with examples when they are encountered so as to forge links with other more familiar objects in the pantry. In this way, more extreme and less familiar examples can come to mind as part of an enriched example-space.

The mark of the expert is knowing what to do with examples and counter-examples. In this chapter we demonstrate some of the fruitful ways of exploiting counter-examples so as to get the most from them.

Case Study: Bigger Values Means Bigger Slope?

These are notes assembled by one of the authors (John Mason) as a case study in constructing counter-examples. The idea is to illustrate some of the ways that counter-examples are constructed, and the playfulness of mathematical exploration which can enrich students' appreciation of concepts.

Consider Conjecture 4.1 from Chapter 3 that if both functions $f(x)$ and $g(x)$ are differentiable on an open interval (a, b) and if at each point x in the interval $f(x) > g(x)$, then the derivatives are similarly related: $f'(x) > g'(x)$ on (a, b).

Casting this in terms of slope, it says that if at every point x, the value of f is greater than the value of g, then at every point x, the slope of f is greater than the slope of g as well. The counter-example offered is two straight lines. Why?

The simplest slope is a constant slope (rather than a slope which is changing as x changes), and the very simplest is a slope of 0. Since this is to be a counter-example, we specify f as having slope 0, and then we look for a second function g whose slope is greater than 0, say 1. The functions $f(x) = 1$ and $g(x) = x$ will serve, as long as the interval (a, b) is to the left of the point $x = 1$

where the two curves cross. If the interval (a, b) has been specified in advance, the function g can be translated so that its graph is always below the graph of f on (a, b).

Since one counter-example defeats the conjecture, the task is completed. However, to stop here would be to miss the power and value of working with counter-examples. Notice first that implicitly we have constructed an infinite class of counter-examples. You can alter the function g by adding suitable constants; you can alter f similarly; you can adjust f and g to be more complicated or sophisticated. You can adjust the functions to cope with any interval (a, b) or indeed $[a, b]$. But a great deal can be gained by probing further. What might other examples look like? Is it possible to characterise all possible counter-examples in some way? To achieve this will be to fully appreciate what is wrong about the conjecture, and hence to appreciate the difference between function values and slopes at points.

Take some function which is familiar and confidence inspiring but not too simple. For example $f(x) = x^2$. What must a function g look like, and over what intervals, in order to produce a counter-example? Strictly we need only one point where the derivatives reverse their order, but if it happens at a single point, it will happen on some subinterval because of the differentiability of both functions. A few moments looking at a sketch of $f(x) = x^2$ suggests that g needs to lie below f but to have a steeper slope. Since the slope of f is given by $f'(x) = 2x$, $g'(x)$ needs to be greater than $2x$ on some subinterval of the specified interval. Suitable slopes might be $2x + 1$, or $3x$ as long as the subinterval is suitably chosen where the functions do not cross, for example, on $[0, 1]$.

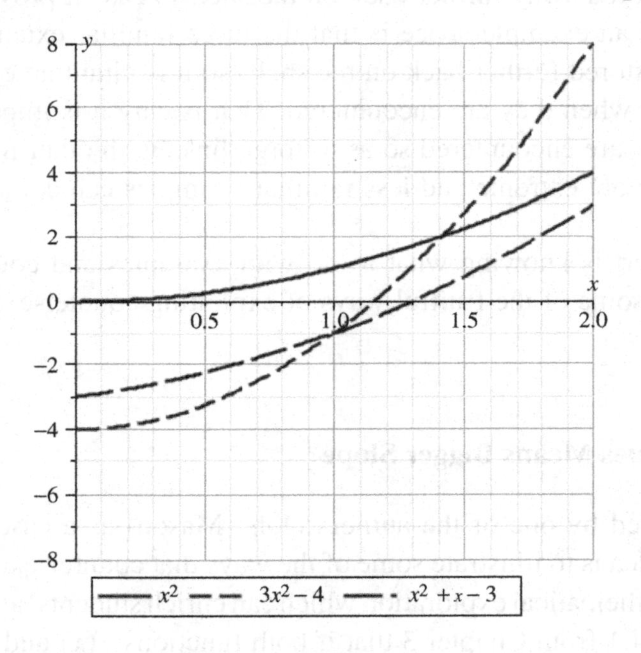

Translating the corresponding g downwards achieves the counter-example. Notice how probing beyond a first simple counter-example provides access not just to more examples, but to a sense of what makes the counter-example work.

Is there anything special about having the functions specified on a finite interval rather than on the whole of R? In other words, are there functions specified on the whole of R which are counter-examples everywhere? A first attempt might be to look for polynomials. Resorting to a familiar example, $f(x) = x^3$ might be a fruitful place to start. It has a slope of $3x^2$ which is always non-negative, so a slope of $6x^2$ would be greater everywhere. This leads to trying $g(x) = 2x^3$. Unfortunately, while working well for $x < 0$ (in the sense that $f > g$ but $f' < g'$), the functions cross at $x = 0$ and then $g > f$ for $x > 0$.

It turns out that the task is impossible when confined solely to polynomials on the whole of R! The polynomials $f(x) = x^3$ and $g(x) = 2x^3$ serve as counter-examples on $(-\infty, 0)$, but there are no polynomials on $(0, \infty)$ that serve as counter-examples on the whole interval. The reason is that if you put $F = f - g$, then you want $F(x) > 0$ but $F'(x) < 0$ on the whole interval, and this is not possible for polynomials.

Before giving up, perhaps it is possible to make use of this function where it does work, and piece it together with something that works on the rest of R. Take $x = 0$ as a glue point. Glue the polynomials to some functions that do work for non-negative x. Moving beyond polynomials, rational polynomials offer more opportunities. When I eventually realised polynomials would not work and tried using reciprocals, I found

$$f(x) = \begin{cases} x^3 + 1 & \text{for } x < 0 \\ x + \dfrac{1}{x+1} & \text{for } x \geq 0 \end{cases} \qquad f'(x) = \begin{cases} 3x^2 & \text{for } x < 0 \\ 1 - \dfrac{1}{(x+1)^2} & \text{for } x \geq 0 \end{cases}$$

$$g(x) = \begin{cases} 2x^3 + x & \text{for } x < 0 \\ x & \text{for } x \geq 0 \end{cases} \qquad g'(x) = \begin{cases} 6x^2 + 1 & \text{for } x < 0 \\ 1 & \text{for } x \geq 0. \end{cases}$$

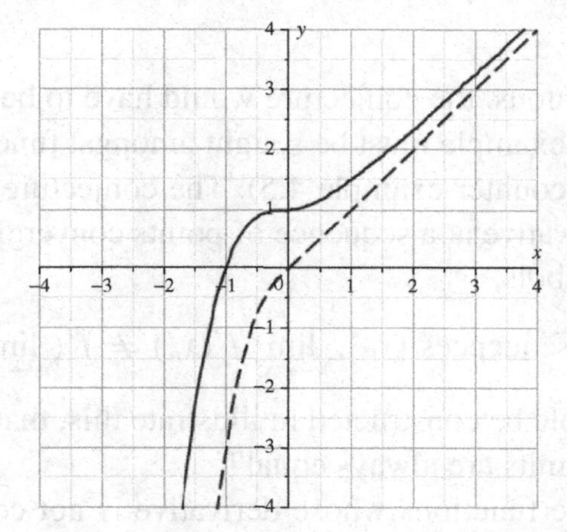

These functions provide a counter-example everywhere on R.

For a slightly different approach, a little algebraic simplification can also help. Let $F(x) = f(x) - g(x)$. Then a counter-example would have $F(x) > 0$ on R, but $F'(x) < 0$ on R. Is this possible? At least the thinking has been reduced to one function rather than trying to cope with two!

If we want to extend to other functions beyond powers and rational polynomials on R, then $f(x) = e^{-x}$ and $g(x) = 0$ work on R, and $f(x) = e^{-2x}$ or $e^{-x} + 2$ or $2e^{-x}$ or 2^{-x} will all work with $g(x) = 0$ as counter-examples.

On Constructing Examples and Counter-Examples

On the face of things, it would seem difficult to construct a counter-example without already appreciating and understanding what the assertion means. Consequently constructing counter-examples might not even be possible, much less relevant, to a student who does not yet understand. This view point was considered by Selden & Selden (1998):

> "If I Don't Know What It Says, How Can I Find an Example of It?"
>
> This hypothetical quote, illustrates the chicken-and-egg quandary some students might typically face when encountering a formal definition, whether of "fine function" or quotient group. A definition asserts the existence

of something having certain properties. However, the student has often never seen or considered such a thing. To give an example or non-example, he/she would need at least some understanding of the concept. But how can he/she obtain such understanding? A good, and possibly the best, way seems to be through an examination of examples. Thus, the student is faced with an epistemological dilemma: Mathematical definitions, by themselves, supply few (psychological) meanings. Meanings derive from properties. Properties, in turn, depend on definitions. [This is a paraphrase from Richard Noss' plenary address to the September 1996 Research in Collegiate Mathematics Education Conference, as reported in *Focus* 17 (1), 1&3, February (1997)].

The next case study highlights ways in which counter-examples arise and are constructed, usually by using examples already encountered, or by tinkering with known examples in a process known as *bricolage*. The study goes beyond mere construction through reflecting on general principles and ways of thinking that have proved fruitful in the past.

Case Study: Limits of Derivatives Are Not Derivatives at Limits

Take for example the conjecture that

If a function is differentiable on $[-1, 1]$ and if its derivative is 0 at $x = 0$ then as x approaches zero, the derivative must also approach 0.

If the derivative were continuous, the conjecture would have to be true. But the derivative need not be continuous so a counter-example must be sought amongst functions whose derivative is not continuous (e.g. $x^{1/3}$ which is counter-example 4.5). The conjecture being false is tantamount to saying that the limit of the derivative at a sequence of points converging to 0 is not necessarily the derivative at the limit 0. In symbols,

$$\text{for some sequences } \{x_n\},\ \lim_{n\to\infty} f'(x_n) \neq f'(\lim_{n\to\infty}(x_n)).$$

How might a counter-example be constructed to illustrate this, that is to act as counter-example to the conjecture that the two limits are always equal?

Contemplating differentiable functions whose derivative is not continuous leads to the observation that the discontinuity cannot be because of a sudden jump in values. A function like

$$s(x) = \begin{cases} 1 & \text{if } x \geq 0 \\ -1 & \text{if } x < 0 \end{cases}$$

cannot be the derivative of any function on an interval which includes 0 because derivatives always satisfy the intermediate value property, and so cannot involve such a jump discontinuity. It does however match the derivative of $|x|$ everywhere except at $x = 0$, where the discontinuity of the derivative of $|x|$ appears.

So what might be useful would be functions which fail to have a derivative at a point for other reasons, such as being undefined (unbounded) there. The functions

$$r_n(x) = \begin{cases} x^{\frac{1}{n}} & \text{if } x \geq 0 \\ -(-x)^{\frac{1}{n}} & \text{if } x < 0 \end{cases}$$

where n is a positive integer, are made by gluing together root-functions for positive x with its rotation through 180° about the origin. They come to mind as tinkerings with the root functions so as to create differentiable functions (apart from at 0). These functions have the property that as n

gets large, they get closer and closer to $s(x)$. Here are $r_n(x)$ for $n = 1$ to 20, and $r_{100}(x)$. They are differentiable everywhere except at $x = 0$, but because the derivative of $r_n(x)$ is

$$\begin{cases} \frac{1}{n}x^{\frac{1}{n}-1} & \text{if } x > 0 \\ \text{undefined} & \text{if } x = 0 \\ \frac{1}{n}(-x)^{\frac{1}{n}-1} & \text{if } x < 0 \end{cases}$$

and for each $n > 1$, the derivative becomes unbounded as x approaches zero.

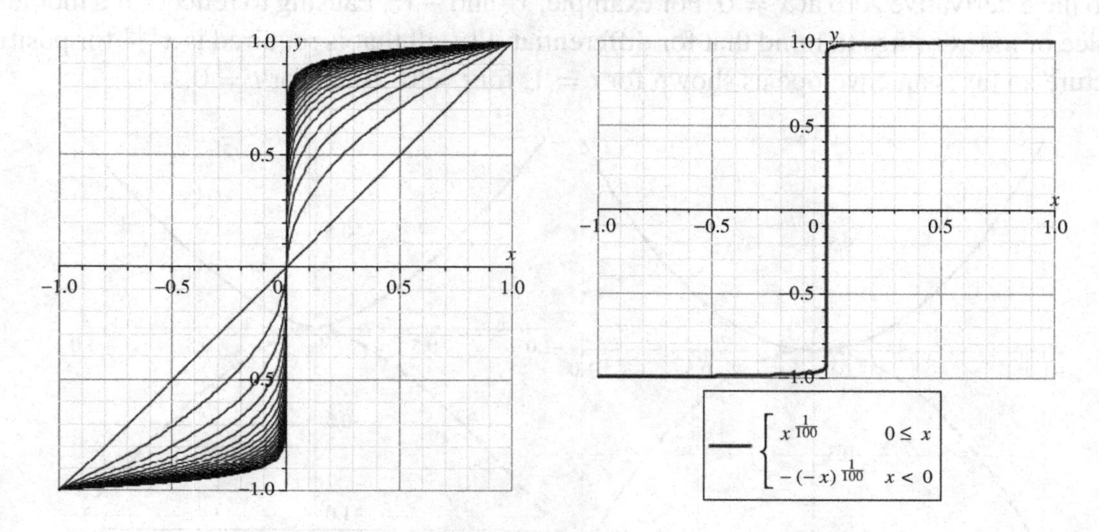

Each of the $r_n(x)$ is differentiable everywhere except at $x = 0$. Integrating the r_n gives the functions

$$R_n(x) = \begin{cases} \frac{n}{n+1}x^{1+\frac{1}{n}} & \text{if } x \geq 0 \\ \frac{n}{n+1}(-x)^{1+\frac{1}{n}} & \text{if } x < 0 \end{cases}$$

whose derivatives are of course the $r_n(x)$. Here is $R_{20}(x)$:

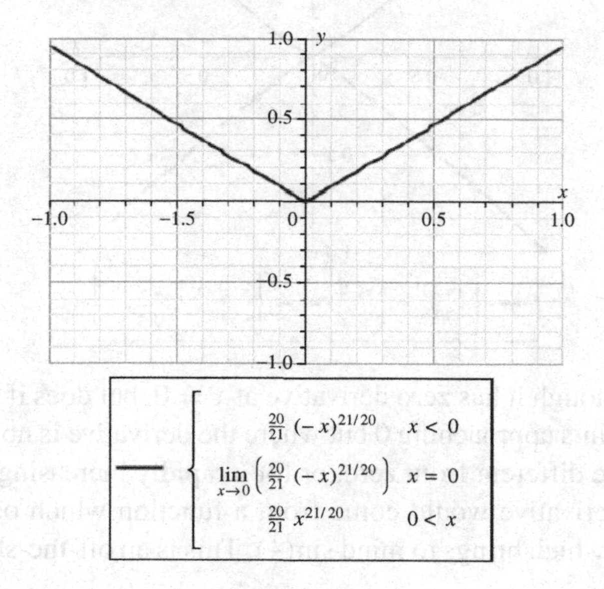

It looks very much like $y = |x|$ on the interval $[-1, 1]$. However its derivative at $x = 0$ is actually 0, because it is ever so slightly rounded near $x = 0$. So we have a family of differentiable functions whose pointwise limit (limit for each value of x) is a function which is not itself differentiable everywhere. This is not exactly what we were looking for, but it illustrates how it is possible to trip over useful functions for other purposes, and how it is worthwhile exploring in order to accumulate examples which might be useful elsewhere. If you keep your attention single-mindedly on the task, you may miss encountering examples to store in your pantry for future use.

Back to the drawing board. In order to get a derivative of 0 at $x = 0$, it is useful to squeeze some function between two differentiable functions both of which have the same value (say 0, at $x = 0$) and also have derivative zero at $x = 0$. For example, x^2 and $-x^2$. Pausing to reflect for a moment on the choice of index brings to mind that for differentiability, all that is required is x^{1+t} for positive t. The picture so far is an envelope as shown for $t = 1$, for $t = 0.5$, and for $t = 0.1$.

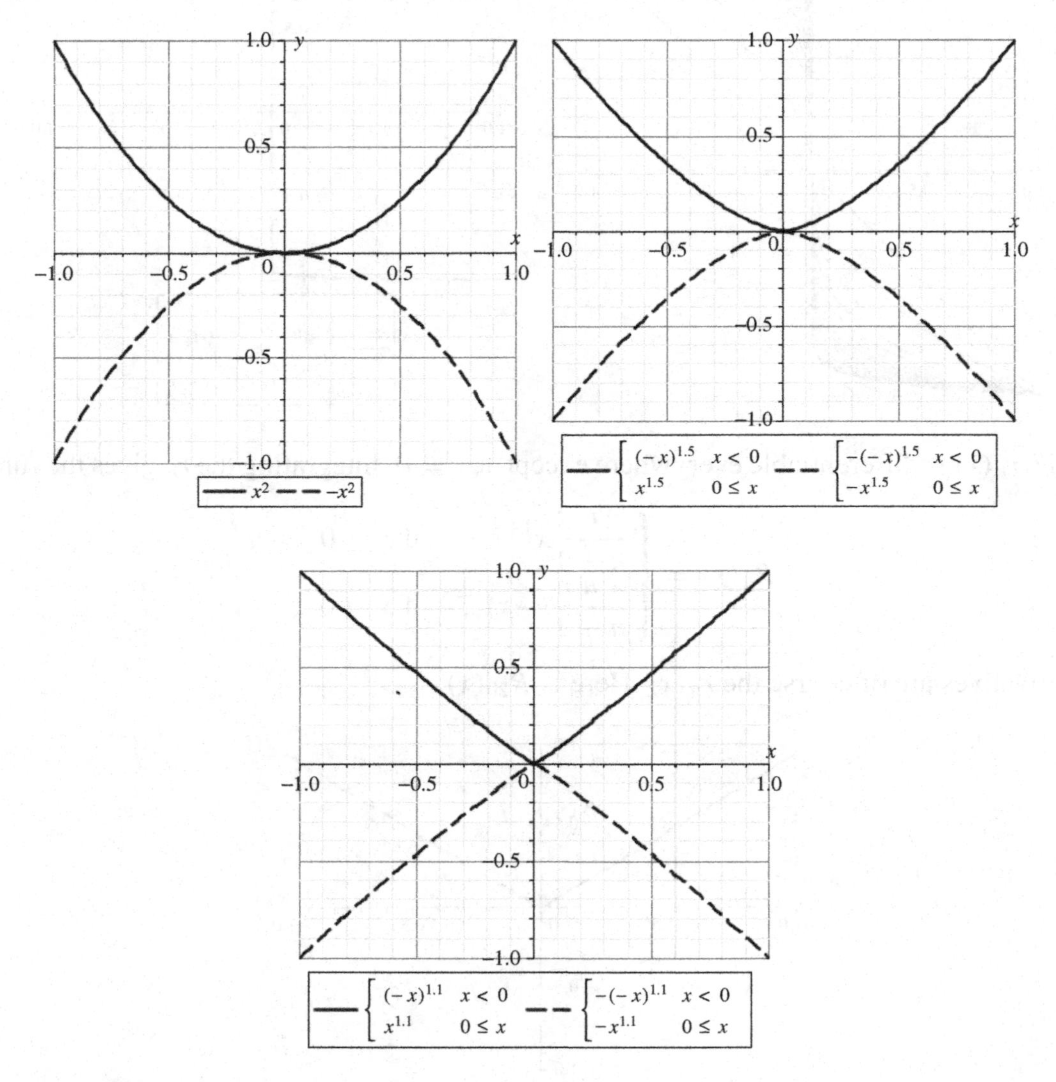

The last does not look as though it has zero derivative at $x = 0$, but does if you zoom in far enough. To create a sequence of points approaching 0 but where the derivative is not approaching 0, you can go for a constant derivative different from zero, or for a rapidly increasing derivative.

A rapidly increasing derivative would come from a function which oscillates more and more rapidly as x approaches 0, which brings to mind $\sin(\frac{1}{x})$. This is an off-the-shelf classic function, just one of an infinite class which oscillates more and more wildly as it approaches $x = 0$. To squeeze it between differentiable functions, consider $x^{1+t}\sin(\frac{2\pi}{x})$. For t positive, this will be differentiable everywhere. At $x = \frac{1}{n}$ it will be 0. Its derivative is $(1+t)x^t\sin(\frac{2\pi}{x}) - \frac{2\pi x^{1+t}}{x^2}\cos(\frac{2\pi}{x})$ which at $x = \frac{1}{n}$ is $0 - 2\pi n^{t-1}$. For t less than 1, $t - 1$ is negative, so n^{t-1} gets smaller and smaller in absolute value as n gets larger. The case of $t = \frac{1}{3}$ is shown in the plot below.

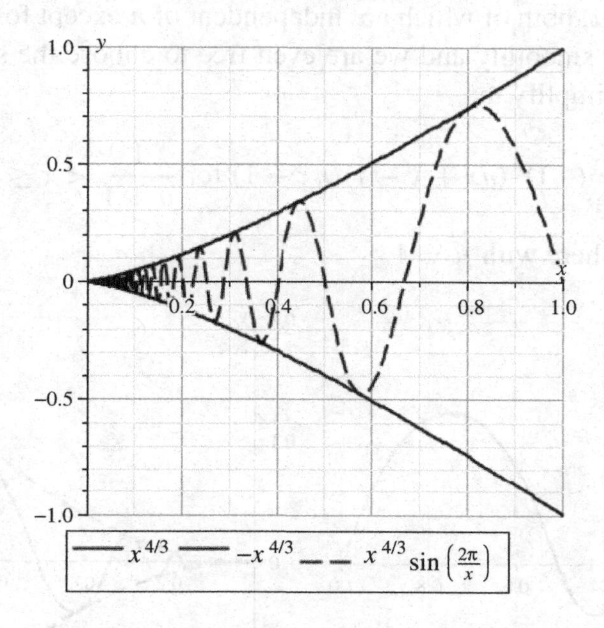

Note that the function is assembled from two familiar pieces, $\sin(\frac{1}{x})$ and x^{1+t}. This is an example of bricolage.

The rest of this case study is a demonstration of what can happen when you allow yourself to be playful, or perhaps self-indulgent. It is much more interesting and satisfying to do the exploring yourself than it is to read the results of someone else's exploration.

Not being satisfied with one class of counter-examples, consider constructing a function for which the derivative is constant on some sequence approaching $x = 0$.

The quadratic $a(x-b)(x-c)$ is zero when $x = b$ and c and has slopes of $\pm a(b-c)$ at b and c respectively. We can therefore glue together a sequence of quadratics which are alternately facing upwards and downwards, so that their slopes match at their zeros. For example, taking the interval $[a, b]$ to be $\left[\frac{1}{n+1}, \frac{1}{n}\right]$ then let $a_n (n = 1, 2, \ldots)$ be positive constants to be determined later, and put

$$f_n = a_n(-1)^n\left(x - \frac{1}{n}\right)\left(x - \frac{1}{n+1}\right)$$

$$= a_n(-1)^n\left(x^2 - \frac{2n+1}{n(n+1)}x + \frac{1}{n(n+1)}\right).$$

For positive coefficients a_n, this is a sequence of quadratics alternately facing upwards and downwards and going through 0 at $x = \frac{1}{n}$ for $n = 1, 2, 3, \ldots$. The slopes of these functions at their zeros are found from the derivative

$$Df_n = a_n(-1)^n\left(2x - \frac{2n+1}{n(n+1)}\right)$$

which takes the value

$$\frac{a_n(-1)^n}{n(n+1)} \text{ at } x = \frac{1}{n} \text{ and } \frac{a_n(-1)^{n+1}}{n(n+1)} \text{ at } x = \frac{1}{n+1}.$$

Consequently, choosing a_n to be $n(n+1)\mu$ will make the slope of f_n at $x = \frac{1}{n}$ take the value $(-1)^n\mu$ while at $\frac{1}{n+1}$ it is $(-1)^{n+1}\mu$, both of which are independent of n except for sign. Consequently the functions all glue together smoothly and we are even free to choose the slope at the joins. When $\mu = 1$ the functions then simplify to:

$$f_n = (-1)^n(nx + x - 1)(nx - 1) \text{ for } \frac{1}{n+1} < x \leq \frac{1}{n}$$

and the first six are shown here with $\mu = 1$.

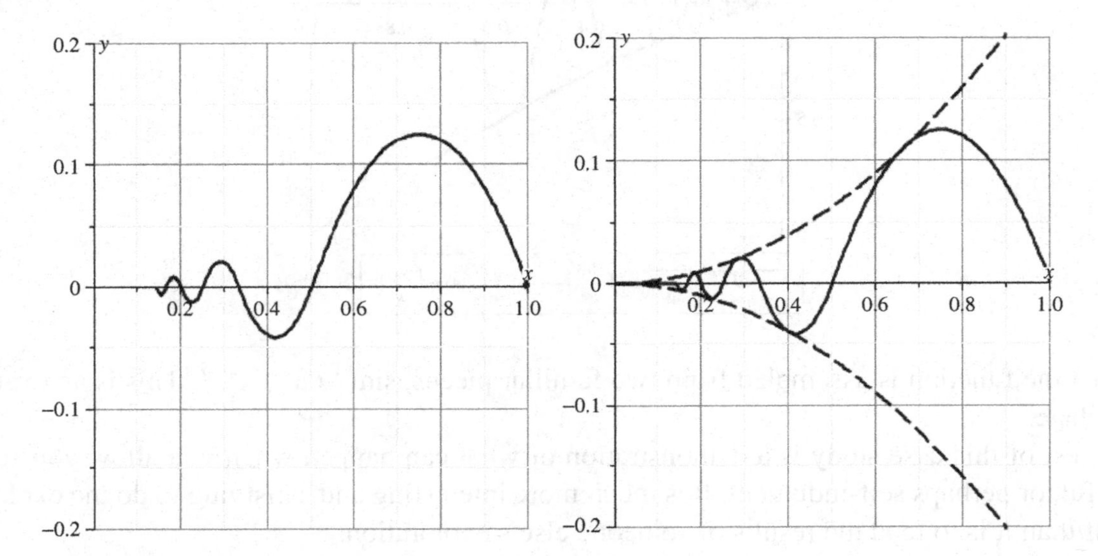

Now it turns out that these functions are all bounded by the functions $\pm\mu\lambda x^2$ as long as $\lambda \geq \frac{1}{4}$ so the compound or glued function must be differentiable at 0 despite having a slope of $\pm\mu$ arbitrarily close to 0. In fact it so happens that the functions $g^+(x) = \frac{\mu x^2}{4}$ and $g^-(x) = \frac{-\mu x^2}{4}$ are tangent to the curves and so bound them tightly.

Thus the function f specified by

$$f(x) = \begin{cases} 0 & \text{if } x = 0 \\ (-1)^n(nx + x - 1)(nx - 1) & \text{if } \frac{1}{n+1} < x \leq \frac{1}{n} \end{cases}$$

is bounded by $\pm x^2/4$, so it is differentiable at $x = 0$. However, at $1/n$ the derivative is always $(-1)^n$. Thus this function is a counter-example to the conjecture.

By reflecting on the construction (making a function oscillate but maintain slopes), the way is open to infinitely many other examples which might be more difficult to state explicitly, but which have similar properties: Oscillating more and more rapidly but with repeating slopes, or as in the case of the first example, slopes getting steeper and steeper as x approaches 0.

Using the same oscillating idea, cubics passing thorough $(\frac{1}{n}, 0)$ and $(1/(n+1), 0)$ and their midpoint can be glued together at the outside zeros so as to have matching slopes there. This produces another class of counter-examples to the conjecture, because again there are points arbitrarily close to 0 at which the derivative is arbitrarily large.

Altering the interval widths on which the quadratics and cubics are assembled is another dimension of possible variation, but it turns out to be problematic! Attempting to use intervals of width $\frac{1}{2}^n$ on which to glue cubics and quadratics runs up against problems: Having matched the derivatives at the end points it is difficult to get unboundedness of the derivative in the interior or else to get differentiability at the limit when $x = 0$. Instead, the curves get flatter and flatter. However, using intervals between reciprocals, everything works out beautifully.

Extending Further

The basic idea is to squeeze an oscillating function which is differentiable almost everywhere, between two functions, both differentiable at the special points. But why not glue together such functions on each interval $[\frac{1}{n+1}, \frac{1}{n}]$? By arranging each to be differentiable with slope 0 at the end points, the result is a function like those constructed above but on each of a countable number of intervals.

First, here is a function which has the oscillations at both ends of the interval [0, 1]:

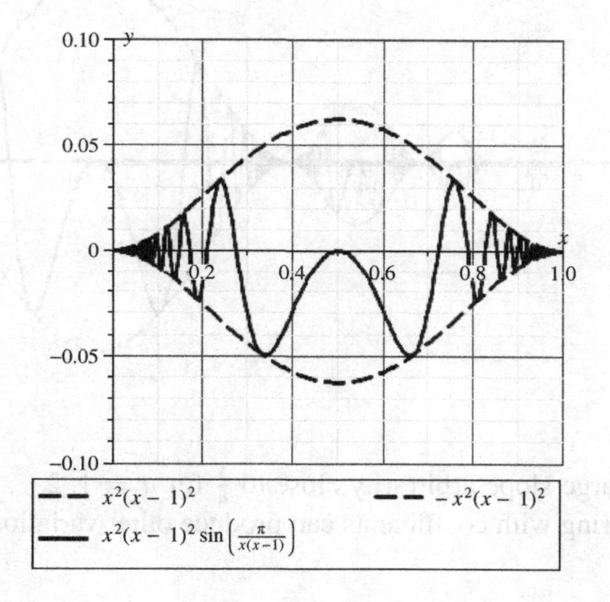

Now the x is scaled so as to have the same appearance but on $[0, \frac{1}{n(n-1)}]$. Then it is translated so as to be on the interval $[\frac{1}{n}, \frac{1}{n} - 1]$:

$$n^2(n-1)^2(nx-1)^2(nx-1-x)^2 \sin\left(\frac{\pi}{n(n-1)(nx-1)(nx-1-x)}\right).$$

Notice that the squares on the outside factor could be replaced by $1 + t$ for positive t and still generate an example. The outside indices form two independent dimensions of possible variation.

The function then looks like the picture on the first when two pieces are glued together on each interval, and on the second, the same function with a scaling of $\frac{1}{n(n-1)}$ on the interval $[\frac{1}{n}, \frac{1}{n-1}]$.

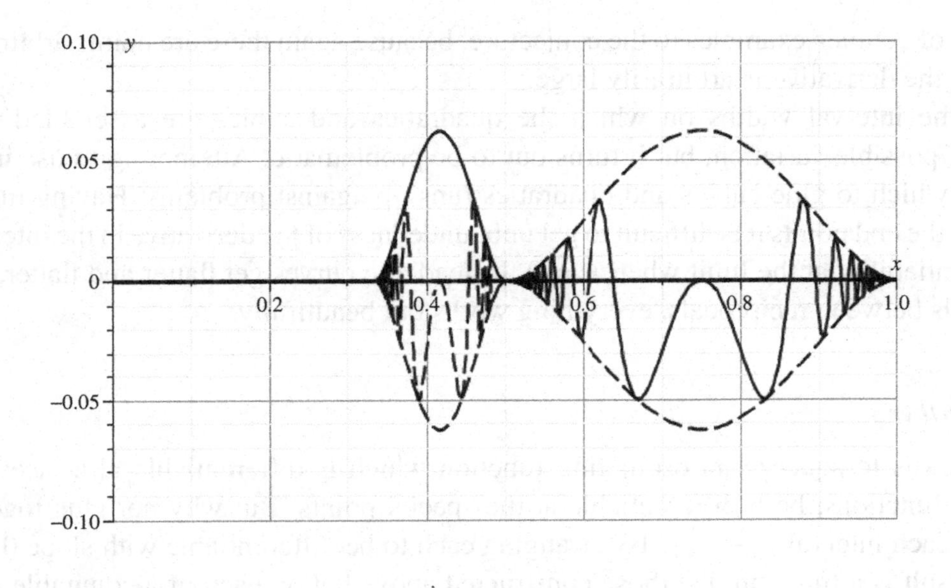

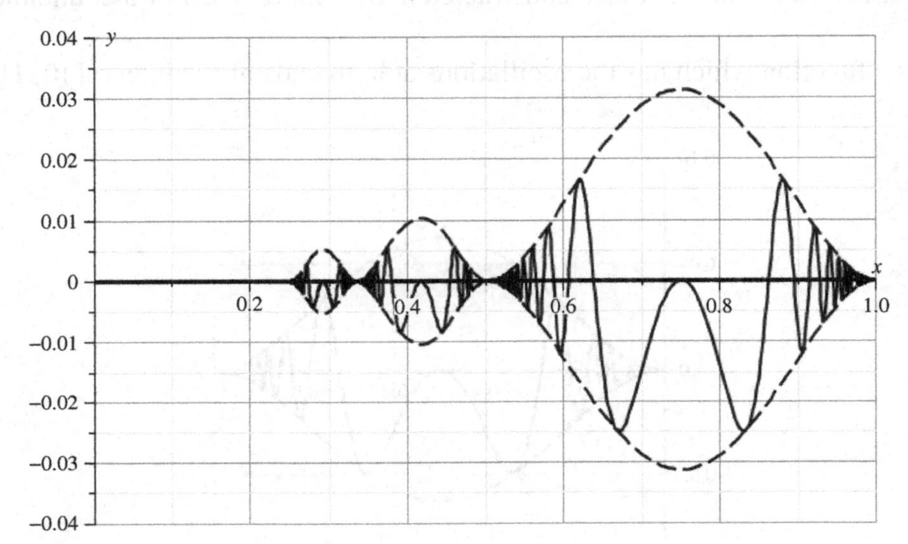

Both have arbitrarily large slope arbitrarily close to $\frac{1}{n}$ for $n = 1, 2, \ldots,$ and are differentiable everywhere. Further tinkering with coefficients can produce other variations.

Useful Strategies

The strategies described in this section have a two-fold purpose. On the one hand they are useful for getting learners to engage deeply with the theorems and techniques they need to encounter, and the underlying concepts. On the other hand they are excellent for illuminating learner misconceptions, inappropriate constraints and assumptions.

Constraint Adjustment

Removing or adjusting a constraint on a true statement and then finding a counter-example to the revised statement not only enriches appreciation of the role of the constraint, but forces learners to examine that role in more detail.

When seeking an example or counter-example meeting several constraints, it is often useful to begin with no constraints at all, and to try to get a sense of the most general class of objects being considered. Constraints can be added sequentially, looking for the most general class of examples that meet one set of conditions before including the next constraint.

Generalising

Extending one or two examples to classes of examples extends learners' appreciation of the scope and significance of the counter-example(s). This counteracts the tendency to isolate and monster-bar (Lakatos, 1976) awkward examples, and so lead learners to appreciate the class of objects admitted as functions. A very useful prompt for extending classes of both examples and counter-examples is to look out for *dimensions-of-possible-variation* (Watson & Mason, 2005): What can be varied and still the object remains an example (or counter-example). Furthermore, with each aspect that can be varied, considering the *range-of-permissible-change* often reveals that learners have a very restricted sense of permissible change.

Starting from a mathematical object (in this case probably a function), learners can be invited to find as many theorems that it illustrates as they can. They can also construct some "conjectures" that others might think are true, to which it is a counter-example. As learners begin to see mathematical functions not as isolated objects but as representatives of classes, their appreciation of theorems improves.

Asking learners to describe, even to characterise all other examples (counter-examples) like one that is given is another way to get them to consider what can be altered or varied, and to what extent, and still preserve the property of being an example (or counter-example).

Bricolage and Tinkering

At first, learners are diffident about tinkering with mathematical objects, and particularly with functions. They tend to see them as unassailable wholes. For example, since gluing functions together caused several famous mathematicians some considerable concern as to whether the result really is a function (their sense had been of a formula rather than a relationship), it is reasonable to expect that learners have similar concerns. Therefore, getting them to glue functions together themselves can support them in gaining a sense of mastery over a space of functions which goes well beyond the polynomials, rational polynomials, exponentials and trigonometric functions. More generally, as Seymour Papert (1993) emphasised, the French notion of *bricolage*, of tinkering with components of familiar objects to make new constructions, is a central component of expertise, and an important contribution to learning.

Asking learners what minimal change can be made to an example to make it into a counter-example, or to a counter-example to make it into an example not only extends learners' appreciation of the conditions and constraints involved, but also contributes to their sense that they have the power to tinker and alter. This in turn enriches their whole sense of example space.

Working with Counter-Examples

Learners can be given a list of statements and a list of functions, and asked which functions are examples, and which counter-examples of which statements. For example, $\frac{1}{2}$ is a counter-example to the conjecture that whole numbers have a whole-number multiplicative inverse; that the mean of two integers is always an integer; that every number has a unique name; that every fraction is represented by an infinite decimal. Similarly $\frac{1}{x}$ on (0, 1) is a counter-example to the conjecture that a continuous function on an open interval attains its maximum; that the integral of a power of x is a power of x, and so on.

Same & Different

When introducing a new concept, or a theorem, learners can be engaged by offering them three or more objects, and asking them to say what is the same and what is different about them. In this way their attention is directed to details which they might otherwise overlook. By carefully choosing the examples to offer, attention can be directed to critical features of the concept or the conditions of the theorem. Very often learners apparent overlooking of conditions and assumptions is because their attention has not previously been directed to make appropriate distinctions (Mason, 2003).

Extending Your Accessible Example Space (the Pantry)

Achieving one counter-example is sufficient to show that the conjecture is false. It may even suggest how the conjecture needs to be modified in order to be correct and justifiable. Extending one example to a class of examples enriches your appreciation of why the conjecture is false and how it needs modifying, as well as shifting from thinking that there are only a few "monsters" that could be barred specifically, to appreciating underlying structural reasons for the modified conjecture. By taking a theorem and finding counter-examples to strengthened versions in which a hypothesis or imposed condition is removed, the necessity of each condition can be appreciated. It is then more likely that you will think to check the conditions and hypotheses before trying to apply the theorem in the future.

By not being satisfied by a single example, even by a class of examples, but by seeking other ways in which a counter-example might be constructed you further enrich your appreciation. You may also encounter some objects which can be stored at the back of the pantry for use in the future, not necessarily directly connected with the initial conjecture. For example, in the case study, despite having been aware of the standard examples previously, it had not come to mind to explore the Dimensions of Possible Variation of the power of x in front of $\sin(\frac{1}{x})$. The family $x^{1+t}\sin(\frac{1}{x})$ was an addition to the example space, and becomes available as an idea to use in other situations. Thus, when constructing the compound function h, the idea came to mind again to extend the class.

Case Study from Teaching Practice

Below are some experiences of using counter-examples with students in a tutorial. Statement 3.14 from Chapter 3 is considered as an example:

> If a function $y = f(x)$ is defined on $[a, b]$ and continuous on (a, b) then for any $N \in (f(a), f(b))$ there is some point $c \in (a, b)$ such that $f(c) = N$.

The only difference between this statement and the Intermediate Value Theorem is that continuity of the function is required only on an open interval (a, b), instead of a closed interval $[a, b]$. In other words, one-sided continuity of the function at the point $x = a$ from the right and at the point $x = b$ from the left is not required. When students were asked to disprove the statement they came up with something like this:

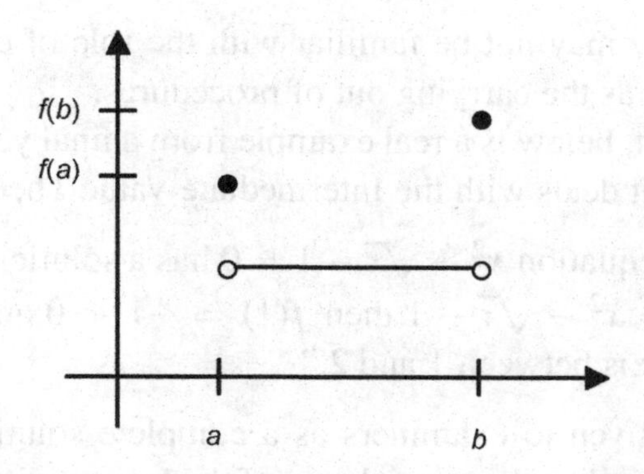

To generate discussion and create other counter-examples they were invited to consider:

In the above graph the statement's conclusion is not true for any value of $N \in (f(a), f(b))$. Modify the graph in such a way that the statement's conclusion is true for:

(a) one value of $N \in (f(a), f(b))$,

(b) infinitely many but not all values of $N \in (f(a), f(b))$.

The students came up with the following sketches:

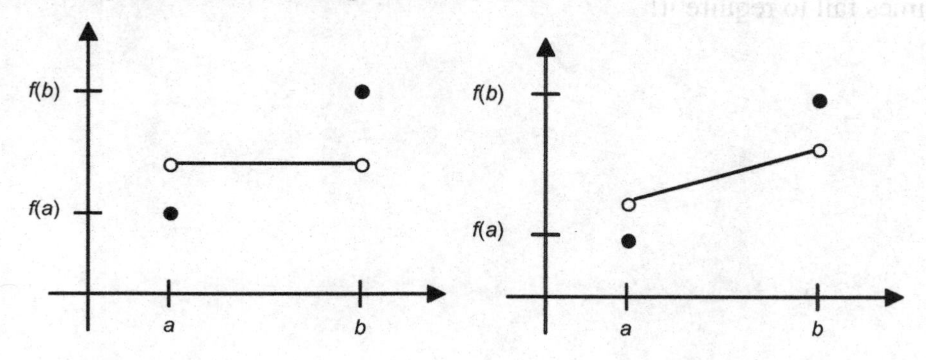

This raises the question of whether it is possible for the conclusion to be true, say for exactly 2 values of N. Another challenge was presented:

Give a counter-example for which the graph does not have white circles.

Eventually students may came up with:

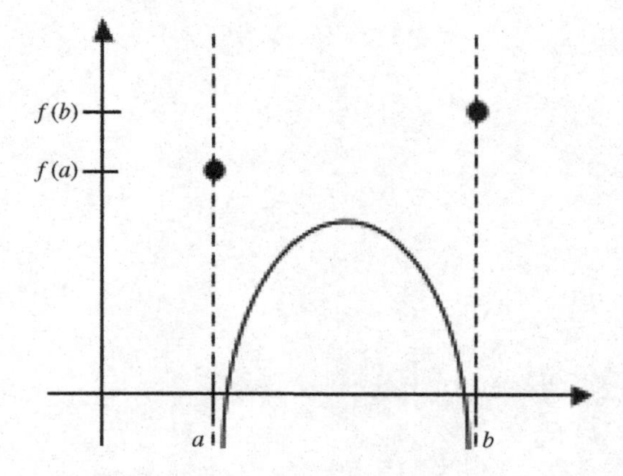

Some students find such problems very new and challenging. After learning calculus at school many come to university with a strong preference for performing calculations, manipulations and techniques, ignoring conditions of the theorems and properties of the functions they are dealing with.

It is often not their fault: They may not be familiar with the role of conditions, and they may be enculturated into mathematics as the carrying out of procedures.

To illustrate the above point, below is a real example from a final year high-school mathematics exam (university entrance) that deals with the Intermediate Value Theorem as well.

> Question. "Show that the equation $x^2 - \sqrt{x} - 1 = 0$ has a solution between $x = 1$ and $x = 2$.
> Model Solution. If $f(x) = x^2 - \sqrt{x} - 1$ then $f(1) = -1 < 0$ and $f(2) = 1.58 > 0$. So the graph of f crosses the x-axis between 1 and 2."

This model solution was given to examiners as a complete solution, one for which students would get full marks. It was based on the special case of the Intermediate Value Theorem which has 2 conditions: The continuity of $f(x)$ on $[a, b]$ and the condition $f(a) \times f(b) < 0$. Only the second condition was checked, and the first was ignored as if it was "not essential". The question came from a written exam where all working had to be shown. The fact that the condition of continuity of the function $f(x)$ was not required by the examiners to award full marks for the solution is very dangerous. The message is clear: Calculations are important but the function's properties are not. No wonder students do not consider all theorem conditions and properties of functions when even experts sometimes fail to require it!

Chapter 2

THE PATHOLOGICAL DEBATE

Each generation of lecturers and teachers re-opens the debate between those who advocate keeping things as simple as possible for students so as not to confuse or overwhelm them, and those who advocate exposing them to "pathological examples" which might extend and enrich their awareness of the scope of definitions, theorems and techniques.

Pro-Pathology

Lecturers who choose to expose students to unusual functions make the sorts of claims that we are making in this book: That if you do not appreciate the scope and range of what functions are possible, the necessary conditions on techniques and theorems will be treated as inessential frills. There are several published accounts of people using formulae and software to design things which have failed because necessary assumptions and conditions were ignored or overlooked.

It may very well turn out that what is unusual now will become commonplace in applications or in new contexts, so it is best to expose students to more than a narrow range of "nice" functions. The last thing we want to have happen is people with only the vaguest notion of what they are doing, applying theorems and techniques inappropriately in contexts where the result really matters.

A third stance is cultural. It argues that the whole enterprise of studying familiar functions, indeed, in studying familiar numbers, is only possible if the scope of enquiry is broadened to all real (or even complex) numbers, all functions, all continuous functions, all differentiable functions, and so on. There is not only a very rich cultural heritage which is valuable for students to experience, but also a significant principle at stake: In order to study one thing and to understand it fully, you may (you will) have to extend your view to include other things as well.

Anti-Pathology

Lecturers who choose not to expose students to unusual functions take the stance that students find things difficult enough without having to cope with the unfamiliar and unusual. Most of these students will never encounter any "nasty" functions, and are only learning calculus as a mathematical basis for some other discipline. They require only the minimum understanding and facility with techniques in order to pass the course. The trouble is that whenever teachers make decisions about what students can and cannot handle, the students respond in kind. Only by challenging students are they likely to rise to that challenge.

The desire to minimise student disturbance, for fear of loss of engagement and motivation usually leads to "dumbing down". This in turn leads to continued student dissatisfaction. Where challenge is reduced or eliminated, students are reaffirmed in their desire to be told only what they really need to "know" in order to pass exams. Where challenge is maintained and expectations are reasonable, students rise to meet them. Furthermore, the more often students can make significant choices for themselves, the more they will enjoy their studies. Example and counter-example construction feeds just such choice making.

Meta-Pathology

Perhaps the disagreement rests in part on the emotive content of the word *pathology*. Mathematicians use the term to indicate examples which are extreme and unexpected. It tends to be used when people come across objects which challenge their intuition. Thus, what is "pathological" at one time or to one group may become familiar and even generative for others. By labelling certain examples as "pathological" we may be doing students a disservice.

In the introduction, under the heading of example spaces, we suggested that impoverished example spaces could be a root cause for students' reluctance to check conditions and hypotheses before applying techniques and theorems in unfamiliar settings. The point was made that if you have only ever encountered "simple" objects, then the entire meaning and purpose of conditions and assumptions may be lost. Thinking and speaking of *example spaces* could bypass the debate entirely and focus attention on more pedagogically relevant issues to do with what it is that students are expected to appreciate about the space of objects to which theorems and techniques apply.

We are by no means the first to promote the use of examples, and example construction:

Alfred North Whitehead

> The progress of science consists in observing . . . interconnexions and in showing with a patient ingenuity that the events of this ever-shifting world are but examples of a few general connexions or relations called laws. To see what is general in what is particular and what is permanent in what is transitory is the aim of scientific thought. (Whitehead, 1911, p. 4)

Paul Halmos

> Let me emphasize one thing . . . the way to begin all teaching is with a question. I try to remember that precept every time I begin to teach a course, and I try even to remember it every time I stand up to give a lecture . . . .
>
> Another part of the idea of the method is to concentrate attention on the definite, concrete, the specific. Once a student understands, really and truly understands, why 3×5 is the same as 5×3, then he quickly gets the automatic but nevertheless exciting and obvious conviction that "it goes the same way" for all other numbers. (Halmos, 1994, p. 852)
>
> The best way to learn is to do; the worst way to teach is to talk. (Halmos, 1975, p. 466)
>
> A good stock of examples, as large as possible, is indispensible for a thorough understanding of any concept, and when I want to learn something new, *I make it my first job to build one*.
>
> . . . Counter-examples are examples too, of course, but they have a bad reputation: They accentuate the negative, they deny not affirm. ... the difference . . . is more a matter of emotion. (Halmos, 1983, p. 63)
>
> If I had to describe my conclusion [as to a method of studying] in one word, I'd say *examples*. They are to me of paramount importance. Every time I learn a new concept I look for examples . . . and non-examples. . . . The

examples should include wherever possible the typical ones and the extreme degenerate ones. (Halmos, 1985, p. 62)

Richard Feynman

I can't understand anything in general unless I'm carrying along in my mind a specific example and watching it go. (Feynman, 1985, p. 244)

George Pòlya

When discussing the issue of whether to try to prove a statement or to find a counter-example, Pòlya suggests a two-pronged approach:

A good scheme [when tackling a problem] is to work alternately, now in one direction, then in the other. When the hope to attain the end in one direction fades, or we get tired of working in that direction, we turn to the other direction, prepared to come back if need be, and so, by learning from our work in both directions, we may eventually succeed. (Pòlya, 1962, pp. 2–51)

He goes on to suggest that sometime you can prove a weaker statement, or even a stronger statement, and sometimes you can disprove a stronger statement (because you have more to work conditions to defeat). He gives an illuminating case study in the construction of counter-examples involving convergent series. (Polya, 1962, pp. 2-49-51)

Annie & John Selden

Since success in mathematics, especially at the advanced undergraduate and graduate levels appears to be associated with the ability to generate examples and counter-examples, what is the best way to develop this ability? One suggestion . . . is to ask students at all levels to "give me an example of . . .". Granted the inherent epistemological difficulties of finding examples for oneself, are we, in a well-intentioned attempt to help students understand newly defined concepts, ultimately hobbling them, by providing them with predigested examples of our own? Are we inadvertently denying students the opportunity to learn to generate examples for themselves? (Selden & Selden, 1998)

Anne Watson & John Mason

Extreme examples, therefore, confound our expectations, encourage us to question beyond our present experience, and prepare us for new conceptual understandings. (Watson & Mason, 2005, p. 7)

By pushing parameters to extremes, surface features can sometimes be confounded (even "multiplication makes bigger" is confounded by multiplication by 1), but sometimes such extreme examples are less rather than more problematic, and certainly not always convincing. Students simply try to "monster bar" such extreme examples away. (Watson & Mason, 2005, p. 11)

If learners are to be adventurous in extending their example spaces, they will inevitably meet the extremes of ranges of permissible change and, hence, bump into non-examples that may at first sight appear to be examples or that demonstrate the importance of qualifying conditions. In other words, working with non-examples helps delineate the example space. Deliberate searching for counter-examples seems an obvious way to understand and appreciate conjectures and properties more deeply. Such a search could be within the current example space or could promote extension beyond. (Watson & Mason, 2007, p. 67)

Sometimes searching for extreme examples (hardest, most complex, or most complicated or tricky) leads to the conclusion that they are all, in fact, easy. This reflects the learners' extension of their example-space and their growing confidence through that extended space. (Watson & Mason, 2005, p. 157)

Chapter 3

BONES TO CHEW: COLLECTION OF FALSE STATEMENTS

This chapter contains incorrect statements from the five major topics found in Introductory Calculus courses: Functions, Limits, Continuity, Differential Calculus and Integral Calculus. The statements from each topic are arranged in order of increasing difficulty. Some statements, especially those in the beginning of each topic, are related to students' regular misconceptions. In the more challenging cases statements often appear to be correct, and many students will be hard-pressed to find counter-examples to them.

3.1. Functions

1.1 The tangent to a curve at a point is the line which touches the curve at that point but does not cross it there.

1.2 The tangent line to a curve at a point cannot touch the curve at infinitely many other points.

1.3 A quadratic function of x is one in which the highest power of x is two.

1.4 If both functions $y = f(x)$ and $y = g(x)$ are continuous and monotone on R then their sum $f(x) + g(x)$ is also monotone on R.

1.5 If both functions $y = f(x)$ and $y = g(x)$ are not monotone on R then their sum $f(x) + g(x)$ is not monotone on R.

1.6 If a function $y = f(x)$ is continuous and decreasing for all positive x and if $f(1)$ is positive, then the function has exactly one root.

1.7 If a function $y = f(x)$ has an inverse function $x = f^{-1}(y)$ on (a, b) then the function $f(x)$ is either increasing or decreasing on (a, b).

1.8 A function $y = f(x)$ is bounded on R if for any $x \in$ R, there is $M > 0$ such that $|f(x)| \leq M$.

1.9 If $g(a) = 0$ then the function $F(x) = \frac{f(x)}{g(x)}$ has a vertical asymptote at the point $x = a$.

1.10 If $g(a) = 0$ then the *rational* function $R(x) = \frac{f(x)}{g(x)}$ (both $f(x)$ and $g(x)$ are polynomials) has a vertical asymptote at the point $x = a$.

1.11 If a function $y = f(x)$ is unbounded and non-negative for all real x, then it cannot have roots x_n such that $x_n \to \infty$ as $n \to \infty$.

1.12 A function $y = f(x)$ defined on $[a, b]$ such that its graph does not contain any pieces of a horizontal straight line cannot take its extreme value infinitely many times on $[a, b]$.

1.13 If a function $y = f(x)$ is continuous and increasing at the point $x = a$ then there is a neighbourhood $(x - \delta, x + \delta)$, $\delta > 0$ where the function is also increasing.

1.14 If a function is not monotone then it does not have an inverse function.

1.15 If a function is not monotone on (a, b) then its square cannot be monotone on (a, b).

3.2. Limits

2.1 If $f(x) < g(x)$ for all $x > 0$ and both $\lim_{x\to\infty} f(x)$ and $\lim_{x\to\infty} g(x)$ exist then

$$\lim_{x\to\infty} f(x) < \lim_{x\to\infty} g(x).$$

2.2 The following definitions of a non-vertical asymptote are equivalent:

(a) The straight line $y = mx + c$ is called a non-vertical asymptote to a curve $f(x)$ as x tends to infinity if $\lim_{x\to\infty}(f(x) - (mx + c)) = 0$.
(b) A straight line is called a non-vertical asymptote to a curve as x tends to infinity if the curve gets closer and closer (as close as we like) to the straight line as x tends to infinity without touching or crossing it.

2.3 The tangent line to a curve at a certain point that touches the curve at infinitely many other points cannot be a non-vertical asymptote to this curve.
2.4 The following definitions of a vertical asymptote are equivalent.

(a) The straight line $x = a$ is called a vertical asymptote for a function $y = f(x)$ if $\lim_{x\to a^+} f(x) = \pm\infty$ or $\lim_{x\to a^-} f(x) = \pm\infty$.
(b) The straight line $x = a$ is called a vertical asymptote for the function $y = f(x)$ if there are infinitely many values of $f(x)$ that can be made arbitrarily large in absolute value as x gets closer to a from either side of a.

2.5 If $\lim_{x\to a} f(x)$ exists and $\lim_{x\to a} g(x)$ does not exist because of oscillation of $g(x)$ near $x = a$ then $\lim_{x\to a}(f(x) \times g(x))$ does not exist.
2.6 If a function $y = f(x)$ is not bounded in any neighbourhood of the point $x = a$ then either $\lim_{x\to a^+} |f(x)| = \infty$ or $\lim_{x\to a^-} |f(x)| = \infty$.
2.7 If a function $y = f(x)$ is continuous for all real x and $\lim_{n\to\infty} f(n) = A$ then $\lim_{x\to\infty} f(x) = A$.

3.3. Continuity

3.1 If the absolute value of the function $y = f(x)$ is continuous on (a, b) then the function is also continuous on (a, b).
3.2 If both functions $y = f(x)$ and $y = g(x)$ are discontinuous at $x = a$ then $f(x) + g(x)$ is also discontinuous at $x = a$.
3.3 If both functions $y = f(x)$ and $y = g(x)$ are discontinuous at $x = a$ then $f(x) \times g(x)$ is also discontinuous at $x = a$.
3.4 A function always has a local maximum between any two local minima.
3.5 For a continuous function there is always a local maximum between any two local minima.
3.6 If a function is defined in a certain neighbourhood of point $x = a$ including the point itself and is increasing on the left from $x = a$ and decreasing on the right from $x = a$, then there is a local maximum at $x = a$.
3.7 If a function is defined on $[a, b]$ and continuous on (a, b) then it takes its extreme values on $[a, b]$.
3.8 Every continuous and bounded function on $(-\infty, \infty)$ takes on its extreme values.

3.9 If a function $y = f(x)$ is continuous on $[a, b]$, the tangent line exists at all points on its graph and $f(a) = f(b)$ then there is a point c in (a, b) such that the tangent line at the point $(c, f(c))$ is horizontal.

3.10 If on the closed interval $[a, b]$ a function is bounded, takes its maximum and minimum values and takes all its values between the maximum and minimum values then this function is continuous on $[a, b]$.

3.11 If on the closed interval $[a, b]$ a function is bounded, takes its maximum and minimum values and takes all its values between the maximum and minimum values then this function is continuous at one or more points or subintervals on $[a, b]$.

3.12 If a function is continuous on $[a, b]$ then it cannot take its absolute maximum or minimum value infinitely many times.

3.13 If a function $y = f(x)$ is defined on $[a, b]$ and $f(a) \times f(b) < 0$ then there is some point $c \in (a, b)$ such that $f(c) = 0$.

3.14 If a function $y = f(x)$ is defined on $[a, b]$ and continuous on (a, b) then for any $N \in (f(a), f(b))$ there is some point $c \in (a, b)$ such that $f(c) = N$.

3.15 If a function is discontinuous at every point in its domain then the square and the absolute value of this function cannot be continuous.

3.16 A function cannot be continuous at only one point in its domain and discontinuous everywhere else.

3.17 A sequence of continuous functions on $[a, b]$ always converges to a continuous function on $[a, b]$.

3.4. Differential Calculus

4.1 If both functions $y = f(x)$ and $y = g(x)$ are differentiable and $f(x) > g(x)$ on the interval (a, b) then $f'(x) > g'(x)$ on (a, b).

4.2 If a non-linear function is differentiable and monotone on $(0, \infty)$ then its derivative is also monotone on $(0, \infty)$.

4.3 If a function is continuous at a point then it is differentiable at that point.

4.4 If a function is continuous on R and the tangent line exists at any point on its graph then the function is differentiable at any point on R.

4.5 If a function is continuous on the interval (a, b) and its graph is a *smooth* curve (no sharp corners) on that interval then the function is differentiable at any point on (a, b).

4.6 If the derivative of a function is zero at a point then the function is neither increasing nor decreasing at this point.

4.7 If a function is differentiable and decreasing on (a, b) then its gradient is negative on (a, b).

4.8 If a function is continuous and decreasing on (a, b) then its gradient is non-positive on (a, b).

4.9 If a function has a positive derivative at every point in its domain then the function is increasing everywhere in its domain.

4.10 If a function $y = f(x)$ is defined on $[a, b]$ and has a local maximum at the point $c \in (a, b)$ then in a sufficiently small neighbourhood of the point $x = c$ the function is increasing on the left and decreasing on the right from $x = c$.

4.11 If a function $y = f(x)$ is differentiable for all real x and $f(0) = f'(0) = 0$ then $f(x) = 0$ for all real x.

4.12 If a function $y = f(x)$ is differentiable on the interval (a, b) and takes both positive and negative values on it then its absolute value $|f(x)|$ is not differentiable at the point(s) where $f(x) = 0$, e.g. $|f(x)| = |x|$ or $|f(x)| = |\sin x|$.

4.13 If both functions $y = f(x)$ and $y = g(x)$ are differentiable on the interval (a, b) and intersect somewhere on (a, b) then the function $\max\{f(x), g(x)\}$ is not differentiable at the point(s) where $f(x) = g(x)$.

4.14 If a function is twice differentiable at a local maximum (minimum) point then its second derivative is negative (positive) at that point.

4.15 If both functions $y = f(x)$ and $y = g(x)$ are non-differentiable at $x = a$ then $f(x) + g(x)$ is also not differentiable at $x = a$.

4.16 If a function $y = f(x)$ is differentiable and a function $y = g(x)$ is not differentiable at $x = a$ then $f(x) \times g(x)$ is not differentiable at $x = a$.

4.17 If both functions $y = f(x)$ and $y = g(x)$ are not differentiable at $x = a$ then $f(x) \times g(x)$ is also not differentiable at $x = a$.

4.18 If a function $y = g(x)$ is differentiable at $x = a$ and a function $y = f(x)$ is not differentiable at $g(a)$ then the function $F(x) = f(g(x))$ is not differentiable at $x = a$.

4.19 If a function $y = g(x)$ is not differentiable at $x = a$ and a function $y = f(x)$ is differentiable at $g(a)$ then the function $F(x) = f(g(x))$ is not differentiable at $x = a$.

4.20 If a function $y = g(x)$ is not differentiable at $x = a$ and a function $y = f(x)$ is not differentiable at $g(a)$ then the function $F(x) = f(g(x))$ is not differentiable at $x = a$.

4.21 If a function $y = f(x)$ is defined on $[a, b]$, differentiable on (a, b) and $f(a) = f(b)$, then there exists a point $c \in (a, b)$ such that $f'(c) = 0$.

4.22 If a function is twice-differentiable in a certain neighbourhood of the point $x = a$ and its second derivative is zero at that point then the point $(a, f(a))$ is a point of inflection for the graph of the function.

4.23 If a function $y = f(x)$ is differentiable at the point $x = a$ and the point $(a, f(a))$ is a point of inflection on the function's graph then the second derivative is zero at that point.

4.24 If both functions $y = f(x)$ and $y = g(x)$ are differentiable on R then to evaluate the limit $\lim_{x\to\infty} \frac{f(x)}{g(x)}$ in the indeterminate form of type $[\frac{\infty}{\infty}]$ we can use the following rule: $\lim_{x\to\infty} \frac{f(x)}{g(x)} = \lim_{x\to\infty} \frac{f'(x)}{g'(x)}$.

4.25 If a function $y = f(x)$ is differentiable on (a, b) and $\lim_{x\to a^+} f'(x) = \infty$ then $\lim_{x\to a^+} f(x) = \infty$.

4.26 If a function $y = f(x)$ is differentiable on $(0, \infty)$ and $\lim_{x\to\infty} f(x)$ exists then $\lim_{x\to\infty} f'(x)$ also exists.

4.27 If a function $y = f(x)$ is differentiable and bounded on $(0, \infty)$ and $\lim_{x\to\infty} f'(x)$ exists then $\lim_{x\to\infty} f(x)$ also exists.

4.28 If a function $y = f(x)$ is differentiable at the point $x = a$ then its derivative is continuous at $x = a$.

4.29 If the derivative of a function $y = f(x)$ is positive at the point $x = a$ then there is a neighbourhood about $x = a$ (no matter how small) where the function is increasing.

4.30 If a function $y = f(x)$ is continuous on (a, b) and has a local maximum at the point $c \in (a, b)$ then in a sufficiently small neighbourhood of the point $x = c$ the function is increasing on the left and decreasing on the right from $x = c$.

4.31 If a function $y = f(x)$ is differentiable at the point $x = a$ then there is a certain neighbourhood of the point $x = a$ where the derivative of the function $y = f(x)$ is bounded.
4.32 If a function $y = f(x)$ at any neighbourhood of the point $x = a$ has points where $f'(x)$ does not exist then $f'(a)$ does not exist.
4.33 A function cannot be differentiable only at one point in its domain and non-differentiable everywhere else in its domain.
4.34 A continuous function cannot be non-differentiable at every point in its domain.
4.35 A function cannot be differentiable at just one point without being continuous in a certain neighbourhood of that point.

3.5. Integral Calculus

5.1 If the function $y = F(x)$ is an antiderivative of a function $y = f(x)$ then $\int_a^b f(x)dx = F(b) - F(a)$.
5.2 If a function $y = f(x)$ is continuous on $[a, b]$ then the area enclosed by the graph of $y = f(x)$, OX, $x = a$ and $x = b$ numerically equals $\int_a^b f(x)dx$.
5.3 If $\int_a^b f(x)dx \geq 0$ then $f(x) \geq 0$ for all $x \in [a, b]$.
5.4 If $y = f(x)$ is a continuous function and k is any constant then: $\int kf(x)dx = k\int f(x)dx$.
5.5 A plane figure of an infinite area rotated about an axis always produces a solid of revolution of infinite volume.
5.6 If a function $y = f(x)$ is defined for any $x \in [a, b]$ and $\int_a^b |f(x)|dx$ exists then $\int_a^b f(x)dx$ exists.
5.7 If neither of the integrals $\int_a^b f(x)dx$ and $\int_a^b g(x)dx$ exist then the integral $\int_a^b (f(x) + g(x))dx$ does not exist.
5.8 If $\lim_{x\to\infty} f(x) = 0$ then $\int_a^\infty f(x)dx$ converges.
5.9 If the integral $\int_a^\infty f(x)dx$ diverges then the function $y = f(x)$ is not bounded.
5.10 If a function $y = f(x)$ is continuous and non-negative for all real x and $\sum_{n=1}^\infty f(n)$ is finite then $\int_1^\infty f(x)dx$ converges.
5.11 If both integrals $\int_a^\infty f(x)dx$ and $\int_a^\infty g(x)dx$ diverge then the integral $\int_a^\infty (f(x) + g(x))dx$ also diverges.
5.12 If a function $y = f(x)$ is continuous and $\int_a^\infty f(x)dx$ converges then $\lim_{x\to\infty} f(x) = 0$.
5.13 If a function $y = f(x)$ is continuous and non-negative and $\int_a^\infty f(x)dx$ converges then $\lim_{x\to\infty} f(x) = 0$.
5.14 If a function $y = f(x)$ is positive and not bounded for all real x then the integral $\int_a^\infty f(x)dx$ diverges.
5.15 If a function $y = f(x)$ is continuous and not bounded for all real x then the integral $\int_a^\infty f(x)dx$ diverges.
5.16 If a function $y = f(x)$ is continuous on $[1, \infty)$ and $\int_1^\infty f(x)dx$ converges then $\int_1^\infty |f(x)|dx$ also converges.
5.17 If the integral $\int_a^\infty f(x)dx$ converges and a function $y = g(x)$ is bounded then the integral $\int_a^\infty f(x)g(x)dx$ converges.

Chapter 4

SUGGESTED SOLUTIONS AND NEW CHALLENGES

This chapter is the collection from the previous chapter augmented by counter-examples with reference to pedagogic aspects which might arise when learners try to construct examples for themselves. Questions are offered to prompt the reader to appreciate general classes of counter-examples, not just particular ones. Comments are offered about how the examples might have arisen, and how they can be tinkered with.

4.1. Functions

Counter-Example 4.1.1

The tangent to a curve at a point is the line which touches the curve at that point but dose not cross it there.

(a) The x-axis is the tangent line to the curve $y = x^3$ but it crosses the curve at the origin.

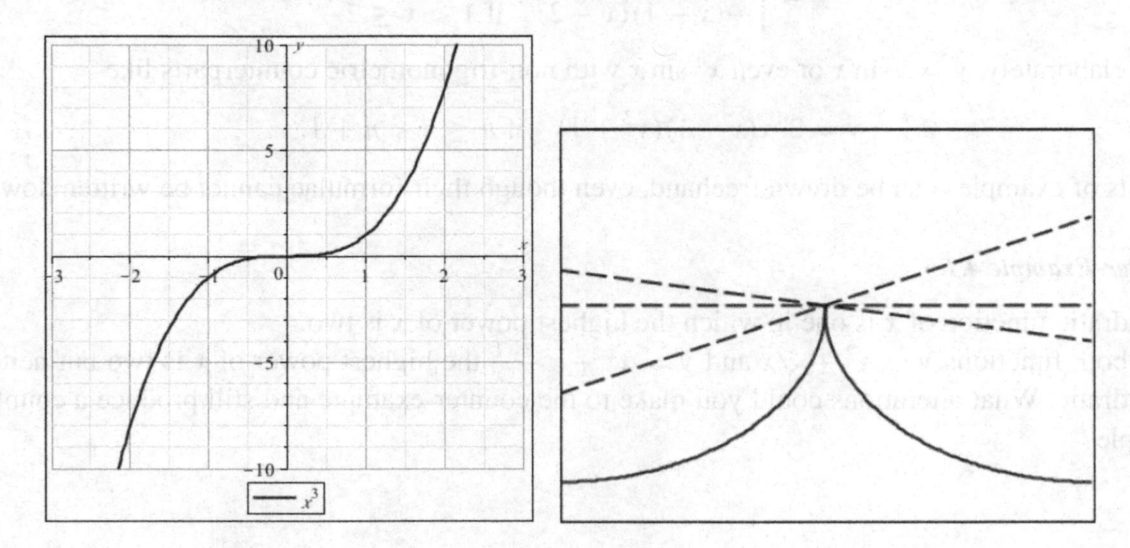

(b) In the second figure, the three straight lines just touch and do not cross the curve at a point but none of them is the tangent line to the curve at that point.

What other functions like $y = x^3$ have a similar "crossing tangent"? Can you make the tangent line which crosses the curve have any specified slope, or must it be horizontal? What other possibilities are there for a cusp point at which there are multiple "touching" lines?

Counter-Example 4.1.2

The tangent line to a curve at a point cannot touch the curve at infinitely many other points.

The tangent line to the graph of the function $y = \sin x$ touches the curve at $x = \frac{\pi}{2}$ and infinitely many other points.

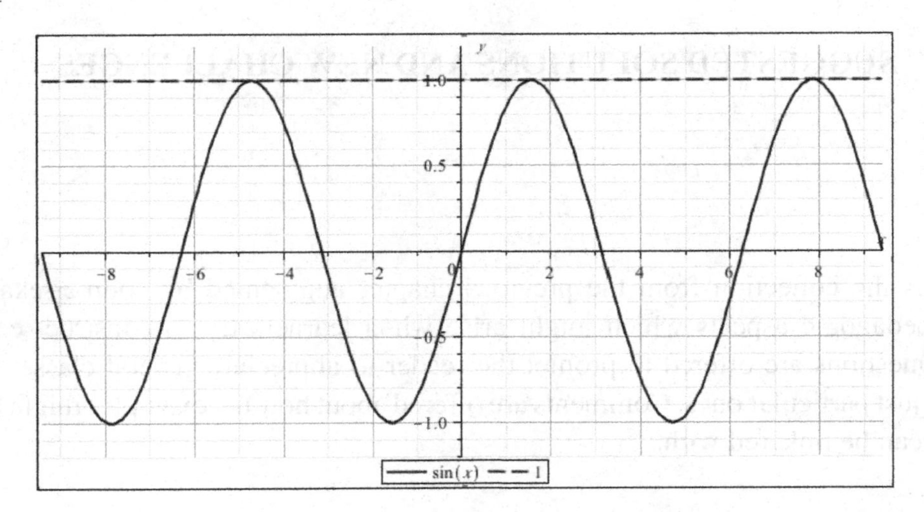

What other functions like $y = \sin x$ can you think of with similar tangents? Can you make up a function which is not based on trigonometric functions? Try assembling a function by gluing together copies of one function, or a class of functions. For example copies of the following function could be translated to the intervals [2, 4], [4, 6], ..., and also into the negatives to produce a periodic function. Notice that the glued function would be differentiable everywhere.

$$y = \begin{cases} x(x-1) & \text{if } 0 \le x \le 1 \\ -(x-1)(x-2) & \text{if } 1 < x \le 2. \end{cases}$$

More elaborately, $y = x \sin x$ or even $x^n \sin x$ with non-trigonometric counterparts like

$$y = 2^n x(x-1)(x^2+1) \quad \text{if } n \le x < n+1.$$

Lots of examples can be drawn freehand, even though their formulae cannot be written down.

Counter-Example 4.1.3

A quadratic function of x is one in which the highest power of x is two.

In both functions $y = x^2 + \sqrt{x}$ and $y = x^2 + x - \frac{1}{x}$ the highest power of x is two but neither is quadratic. What alterations could you make to the counter-example and still produce a counter-example?

Counter-Example 4.1.4

If both functions $y = f(x)$ and $y = g(x)$ are continuous and monotone on R then their sum $f(x) + g(x)$ is also monotone on R.

$$f(x) = x + \sin x, \quad g(x) = -x.$$

Both functions $f(x)$ and $g(x)$ are monotone on R but their sum $f(x) + g(x) = \sin x$ is not monotone on R.

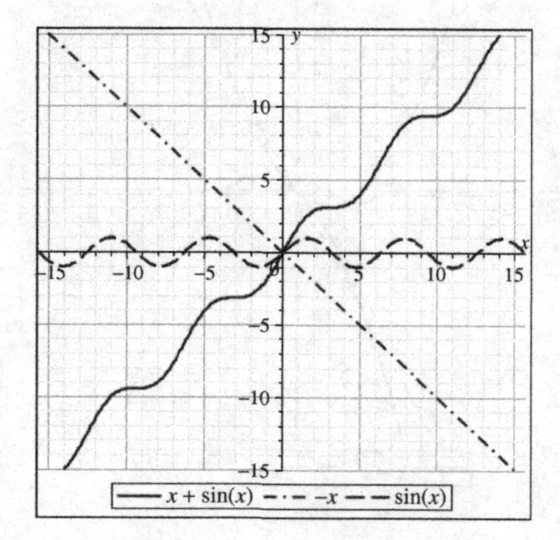

What is it about the example offered which makes it work? Perhaps you thought of monotone as increasing, rather than as either non-decreasing or non-increasing? How could you exploit the same idea with other functions? The function $y = k$ (constant) is also monotone: How could you make use of this to produce a counter-example? For example, you can use zigzag functions such as

$$f(x) = \begin{cases} x + 2 & \text{for } x < -1 \\ -x & \text{for } -1 \leq x \leq 1 \\ x - 2 & \text{for } 1 < x \end{cases}$$

and write them as the sum of two monotone functions.

How obscure can you make two functions and yet have their sum monotone?

Counter-Example 4.1.5

If both functions $y = f(x)$ and $y = g(x)$ are not monotone on R then their sum $f(x) + g(x)$ is not monotone on R.

Both functions $f(x) = x + x^2$ and $g(x) = x - x^2$ are not monotone on R but their sum $f(x) + g(x) = 2x$ is monotone on R.

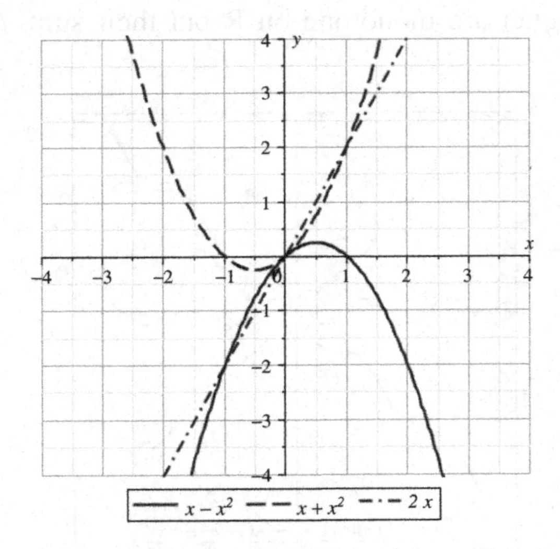

What is the relation between this statement and statement 1.4? In what ways are the examples used to construct the counter-example the same, and in what ways different? Can you exploit a constant function to make a counter-example? What is the most general counter-example you can construct? Did you think of using Pythagoras (e.g. $x\sin^2(x)$ and $x\cos^2(x)$)?

Counter-Example 4.1.6

If a function $y = f(x)$ is continuous and decreasing for all positive x and if $f(1)$ is positive, then the function has exactly one root.

The function $y = \frac{1}{x}$ is continuous and decreasing for all positive x and $y(1) = 1 > 0$ but has no roots.

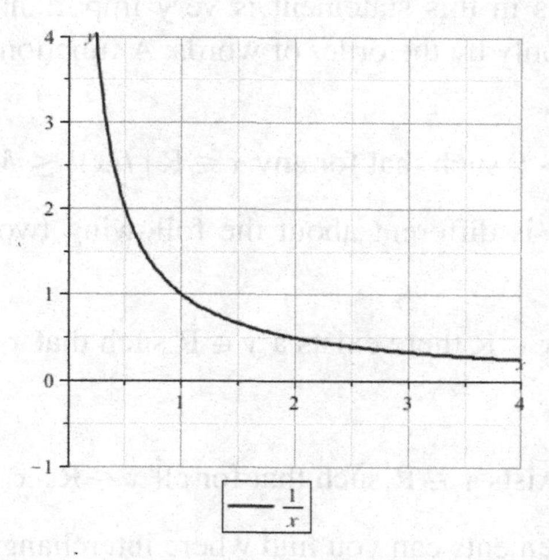

What is it that makes the counter-example work? Was that a surprise? Why can you not make an example which has two roots when $x > 0$? What other functions can you construct which behave like $y = \frac{1}{x}$? What about adding such functions together? Can you make one which oscillates but still fails to cross the x-axis?

Counter-Example 4.1.7

If a function $y = f(x)$ has an inverse function $x = f^{-1}(y)$ on (a, b) then the function $f(x)$ is either increasing or decreasing on (a, b).

The function below is a one-to-one function and has an inverse function on (0, 3) but it is neither increasing nor decreasing on that interval.

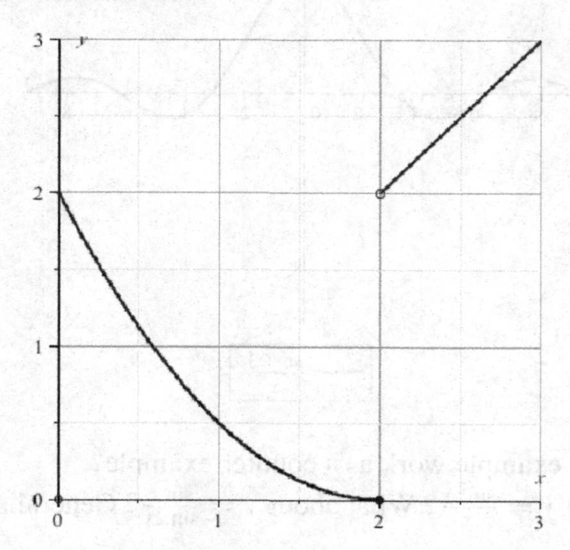

What is it about the counter-example offered which makes it work? Did you assume it had to be continuous? How complicated an example can you make using the same idea?

Counter-Example 4.1.8

A function $y = f(x)$ is bounded on R if for any $x \in$ R there is a number $M > 0$ such that $|f(x)| \leq M$.

For the function $y = x^2$, for any value of x chosen in R, there is a number $M > 0$ ($M = x^2 + \varepsilon$ where $\varepsilon \geq 0$) such that $|f(x)| \leq M$.

Comments. The order of words in this statement is very important. The correct definition of a function bounded on R differs only by the order of words: A function $y = f(x)$ is bounded on R if there is

$$M > 0 \text{ such that for any } x \in \mathrm{R}\ |f(x)| \leq M.$$

What is the same and what is different about the following two statements and the original statements?

$$\text{For all } x \in \mathrm{R} \text{ there exists a } y \in \mathrm{R} \text{ such that } x < y$$

and

$$\text{There exists } y \in \mathrm{R} \text{ such that for all } x \in \mathrm{R},\ x < y.$$

What other examples of statements can you find where interchanging the order of the existence and the for-all makes a difference?

Counter-Example 4.1.9

If $g(a) = 0$ then the function $F(x) = \frac{f(x)}{g(x)}$ has a vertical asymptote at the point $x = a$.

The function $y = \frac{\sin x}{x}$ does not have a vertical asymptote at the point $x = 0$.

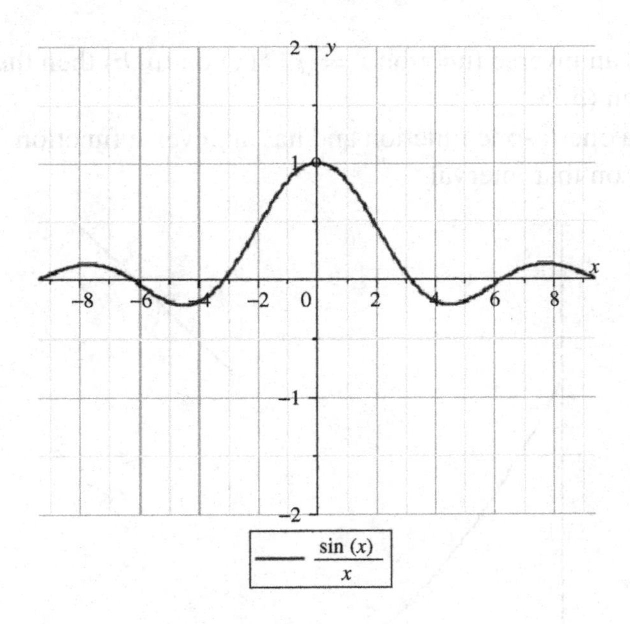

What is it that makes the example work as a counter-example?

What about $y = \frac{\sin x}{x^2}$ and $y = \frac{\sin^2 x}{x}$? What about $y = \frac{\sin x}{\sin 2x}$? Generalise!

Counter-Example 4.1.10

If $g(a) = 0$ then the *rational* function $R(x) = \frac{f(x)}{g(x)}$ (both $f(x)$ and $g(x)$ are polynomials) has a vertical asymptote at the point $x = a$.

The rational function $y = \frac{x^2-1}{x-1}$ does not have a vertical asymptote at the point $x = 1$.

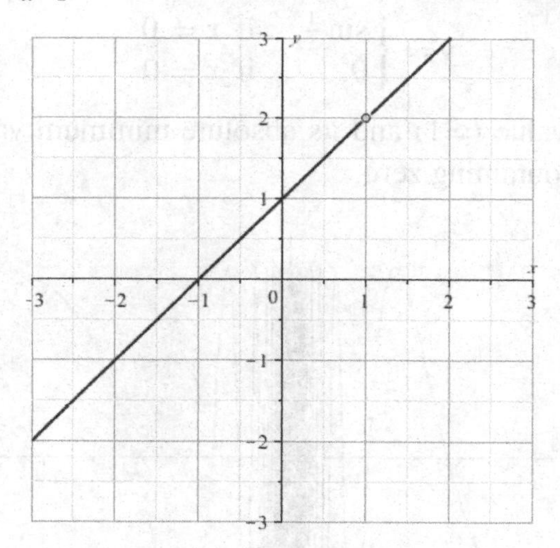

The function given does not even have a value at $x = 1$, though it is very like $y = x + 1$. How could the function be altered so that the value of x of interest in the statement was $x = a$? How could the function be altered so as to be a rational polynomial and not one which is an ordinary polynomial in disguise?

Specifying a value at $x = 1$ other than 2 makes the given example discontinuous. Specifying a value of 2 at $x = 1$ makes it continuous. Generalise to other rational functions which are continuous at the value of x you are considering.

Counter-Example 4.1.11

If a function $y = f(x)$ is unbounded and non-negative for all real x then it cannot have roots x_n such that $x_n \to \infty$ as $n \to \infty$.

The function $y = |x \sin x|$ has infinitely many roots x_n such that $x_n \to \infty$ as $n \to \infty$.

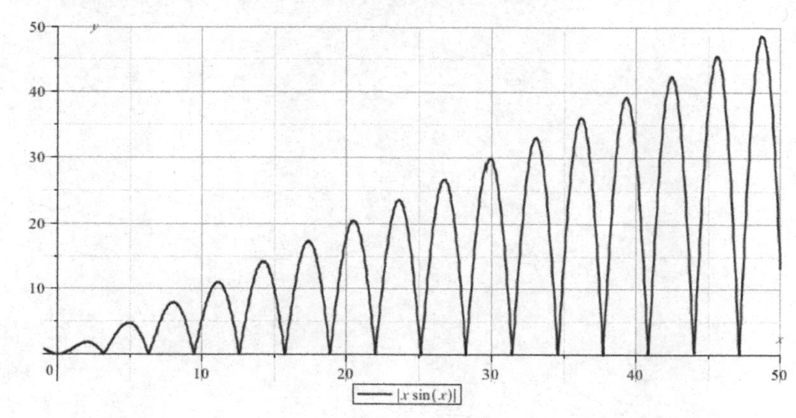

What role does the absolute value play in the example, in terms of being a counter-example to the statement? What other functions could you use as counter-example? What would be the effect on the zeros of using $y = |x \sin(x^2)|$?

Counter-Example 4.1.12

A function $y = f(x)$ defined on $[a, b]$ such that its graph does not contain any pieces of a horizontal straight line cannot take its extreme value infinitely many times on $[a, b]$.

The function

$$y = \begin{cases} \sin\frac{1}{x}, & \text{if } x \neq 0 \\ 0, & \text{if } x = 0 \end{cases}$$

takes its absolute maximum value (=1) and its absolute minimum value (=−1) infinitely many times on any closed interval containing zero.

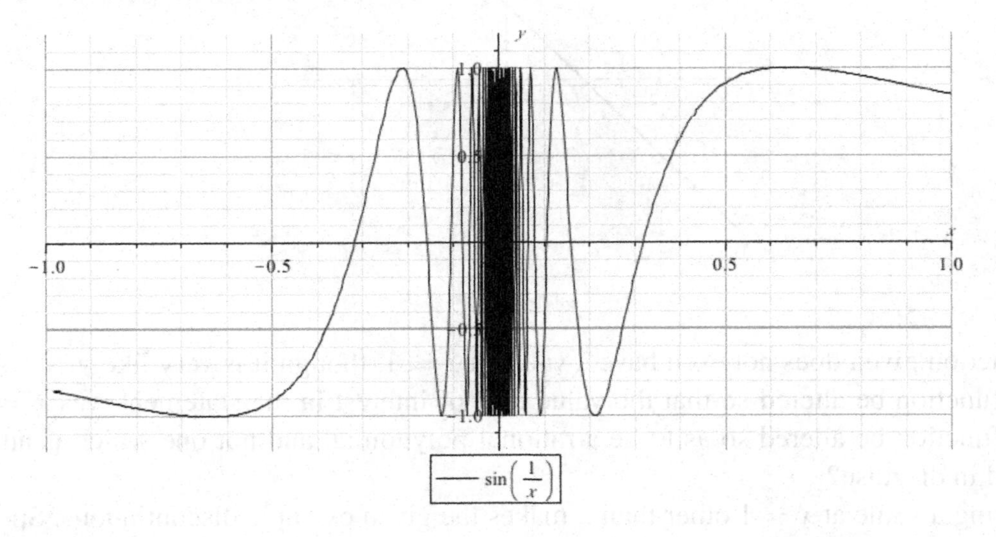

Counter-Example 4.1.13

If a function $y = f(x)$ is continuous and increasing at the point $x = a$ then there is a neighbourhood $(x - \delta, x + \delta)$, $\delta > 0$ where the function is also increasing.

The function

$$f(x) = \begin{cases} x + x^2 \sin \dfrac{2}{x}, & \text{if } x \neq 0 \\ 0, & \text{if } x = 0 \end{cases}$$

is increasing at the point $x = 0$ but it is not increasing in any neighbourhood $(-\delta, \delta)$, where $\delta > 0$.

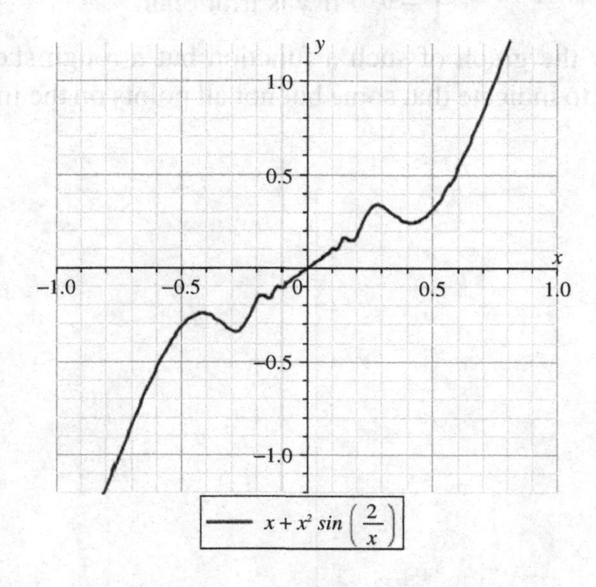

Comments. The definition of a function increasing at a point is:

A function $y = f(x)$ is said to be *increasing at the point* $x = a$ if in a certain neighbourhood $(a - \delta, a + \delta)$, $\delta > 0$ the following is true:

$$\text{if } x < a \text{ then } f(x) < f(a) \text{ and if } x > a \text{ then } f(x) > f(a).$$

Counter-Example 4.1.14

If a function is not monotone then it does not have an inverse function.

The function

$$y = \begin{cases} x, & \text{if } x \text{ is rational} \\ -x, & \text{if } x \text{ is irrational} \end{cases}$$

is not monotone but it has the inverse function

$$x = \begin{cases} y, & \text{if } y \text{ is rational} \\ -y, & \text{if } y \text{ is irrational.} \end{cases}$$

It is impossible to draw the graph of such a function but a rough sketch gives an idea of its behaviour. We use fine dots to indicate that some but not all points on the implied curve are actually on the graph:

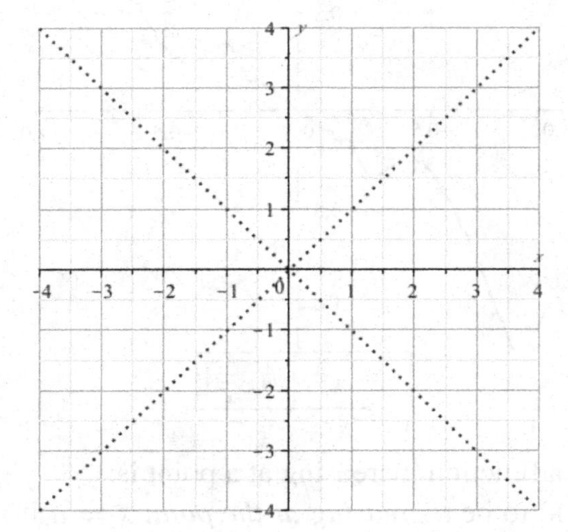

Counter-Example 4.1.15

If a function is not monotone on (a, b) then its square cannot be monotone on (a, b).

The function

$$f(x) = \begin{cases} x, & \text{if } x \text{ is rational} \\ -x, & \text{if } x \text{ is irrational} \end{cases}$$

defined on $(0, \infty)$ is not monotone but its square $f^2(x) = x^2$ is monotone on $(0, \infty)$.

It is impossible to draw the graph of the function $y = f(x)$ but the sketch below gives an idea of its behaviour.

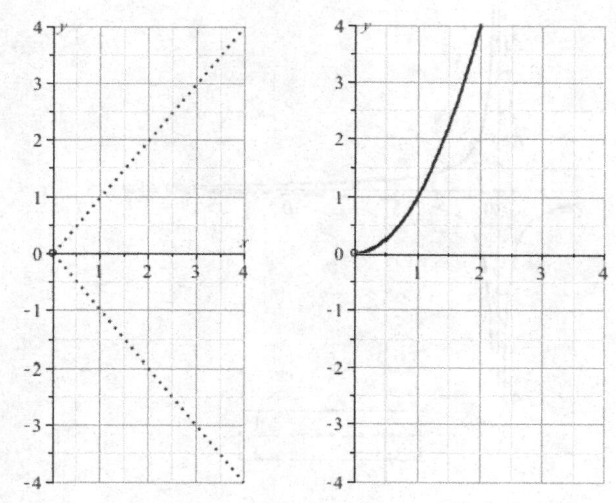

Comments. The functions in counter-examples 4.1.14 and 4.1.15 may seem artificial and without practical use at first. Nevertheless, the Dirichlet function

$$f(x) = \begin{cases} 1, & \text{if } x \text{ is rational} \\ 0, & \text{if } x \text{ is irrational} \end{cases}$$

which is very similar to the functions in counter-examples 4.1.14 and 4.1.15, can be represented analytically as a limit of cosine functions that have many practical applications:

$$f(x) = \lim_{k\to\infty} \lim_{n\to\infty} (\cos(k!\pi x))^{2n}.$$

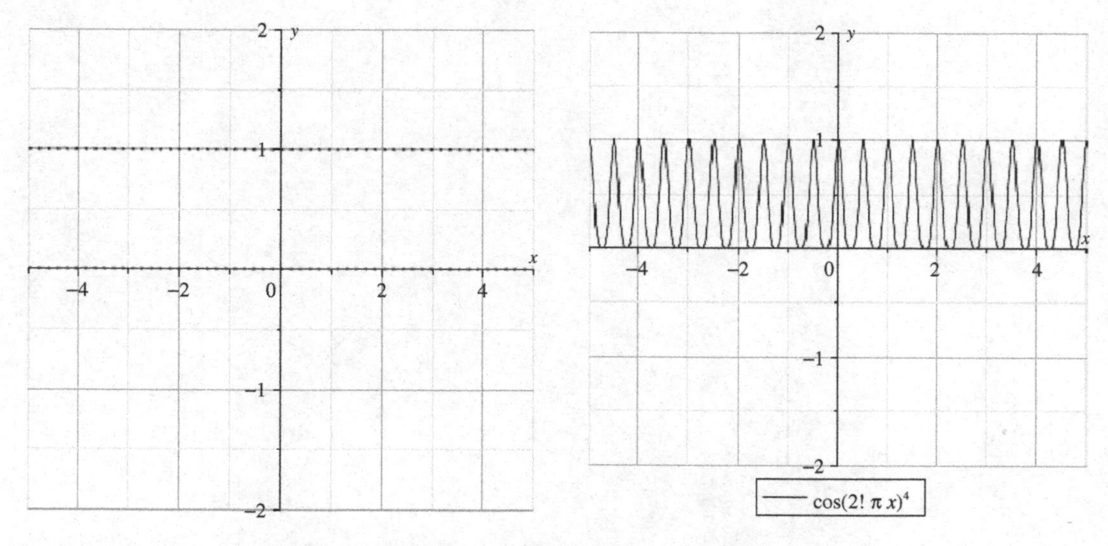

For k and n much bigger than 2, the graph becomes too dense to draw!

4.2. Limits

Counter-Example 4.2.1

If $f(x) < g(x)$ for all $x > 0$ and both $\lim_{x\to\infty} f(x)$ and $\lim_{x\to\infty} g(x)$ exist then $\lim_{x\to\infty} f(x) < \lim_{x\to\infty} g(x)$.

For the functions $f(x) = -\frac{1}{x}$ and $g(x) = \frac{1}{x}$, $f(x) < g(x)$ for all $x > 0$ but $\lim_{x\to\infty} f(x) = \lim_{x\to\infty} g(x) = 0$.

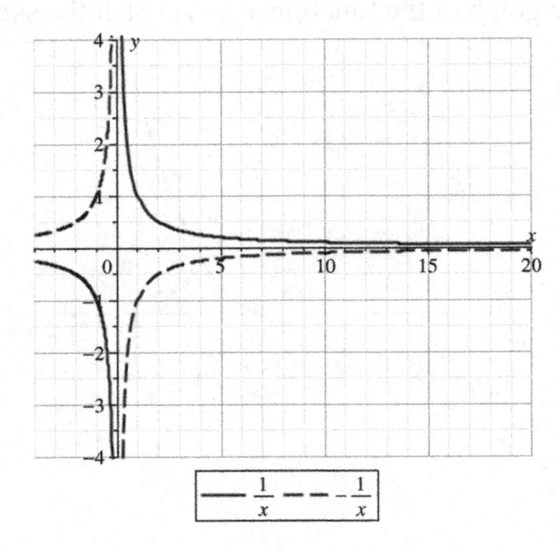

What is it that makes this example work as a counter-example? How could the conjecture be modified to make it correct?

Counter-Example 4.2.2

The following definitions of a non-vertical asymptote are equivalent:

(a) The straight line $y = mx + c$ is called a non-vertical asymptote to a curve $f(x)$ as x tends to infinity if $\lim_{x\to\infty}(f(x) - (mx + c)) = 0$.
(b) A straight line is called a non-vertical asymptote to a curve as x tends to infinity if the curve gets closer and closer to the straight line (as close as we like) as x tends to infinity but does not touch or cross it.

As x tends to infinity the function $y = \frac{\sin x}{x}$ gets closer to the x-axis from above and below and $\lim_{x\to\infty}(\frac{\sin x}{x} - 0) = 0$. According to the first definition the x-axis is the non-vertical asymptote of the function $y = \frac{\sin x}{x}$, but its graph crosses the x-axis infinitely many times, so the definitions (a) and (b) are not equivalent.

Comments. The correct definition is (a). The idea of an asymptotic behaviour is getting closer to a (non-vertical) straight line but this does not exclude touching or crossing it. Note that the function has not been specified at $x = 0$.

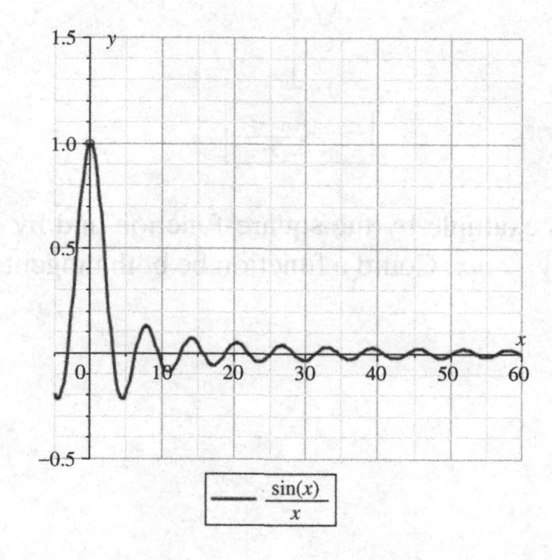

How could the example be modified to use $y = mx$ as the asymptote for $m \neq 0$?

Counter-Example 4.2.3

The tangent line to a curve at a certain point that touches the curve at infinitely many other points cannot be a non-vertical asymptote to this curve.

The tangent line $y = 0$ to the curve $y = \frac{\sin^2 x}{x}$ at $x = \pi$ touches the curve at infinitely many other points and is a non-vertical asymptote to this curve. Note that the function has not been specified at $x = 0$.

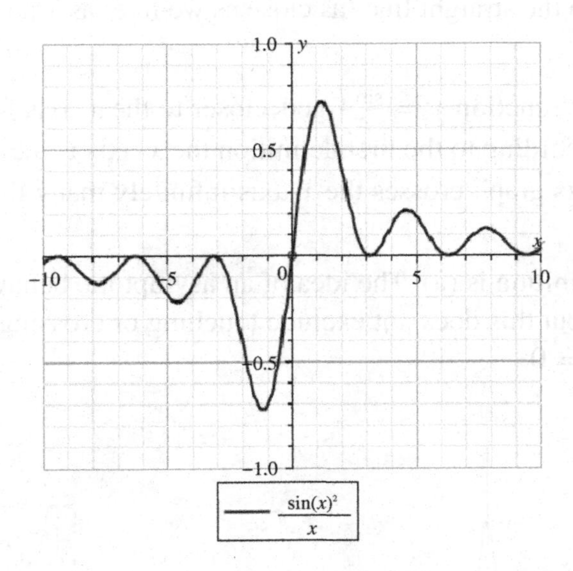

What role is played in this example by the square-function and by $\frac{1}{x}$? Modify the example so that the asymptote is any line $y = mx$. Could a function be both tangent to and cross the asymptote infinitely often?

Counter-Example 4.2.4

The following definitions of a vertical asymptote are equivalent:

(a) The straight line $x = a$ is called a vertical asymptote for a function $y = f(x)$ if $\lim_{x\to a^+} f(x) = \pm\infty$ or $\lim_{x\to a^-} f(x) = \pm\infty$.
(b) The straight line $x = a$ is called a vertical asymptote for the function $y = f(x)$ if there are infinitely many values of $f(x)$ that can be made arbitrarily large in absolute value as x gets closer to a from either side of a.

There are infinitely many values of the function $y = \frac{1}{x} \sin \frac{1}{x}$ that can be made arbitrarily large in absolute value as x gets closer to 0 (for example, when $x = \frac{2}{(2n+1)\pi}$) but the straight line $x = 0$ is not a vertical asymptote of this function because there are also places arbitrarily close to 0 where the value is not large in absolute value (for example when $x = \frac{1}{n\pi}$).

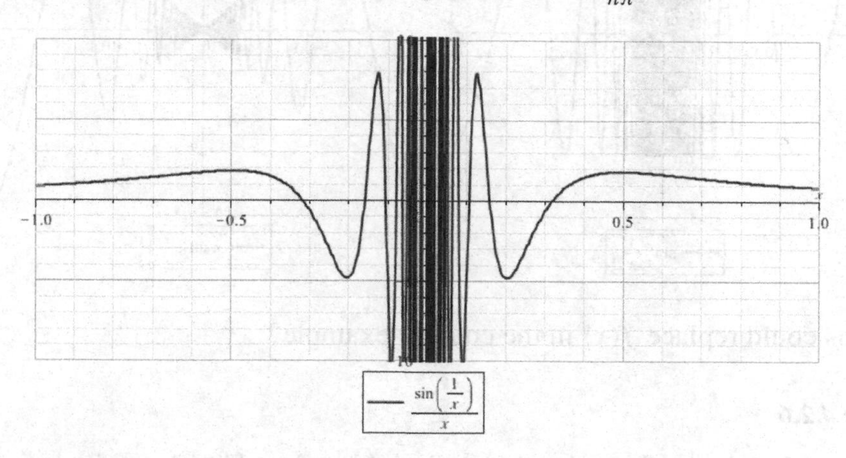

Comments. The correct definition is (a).

What is it about the example which makes it a counter-example? Must all counter-examples oscillate infinitely often? Can you construct a function with finitely many (infinitely many) vertical asymptotes?

Counter-Example 4.2.5

If $\lim_{x\to a} f(x)$ exists and $\lim_{x\to a} g(x)$ does not exist because of oscillation of $g(x)$ near $x = a$ then $\lim_{x\to a}(f(x) \times g(x))$ does not exist.

For the function $f(x) = x$ the limit $\lim_{x\to 0} x = 0$ and for the function $g(x) = \sin\frac{1}{x}$ the limit $\lim_{x\to 0} \sin\frac{1}{x}$ does not exist because of increasingly rapid oscillation of $g(x)$ near $x = 0$, but $\lim_{x\to 0}(f(x) \times g(x)) = \lim_{x\to 0}(x \sin\frac{1}{x}) = 0$. Note that in the second graph, the function is not specified at $x = 0$.

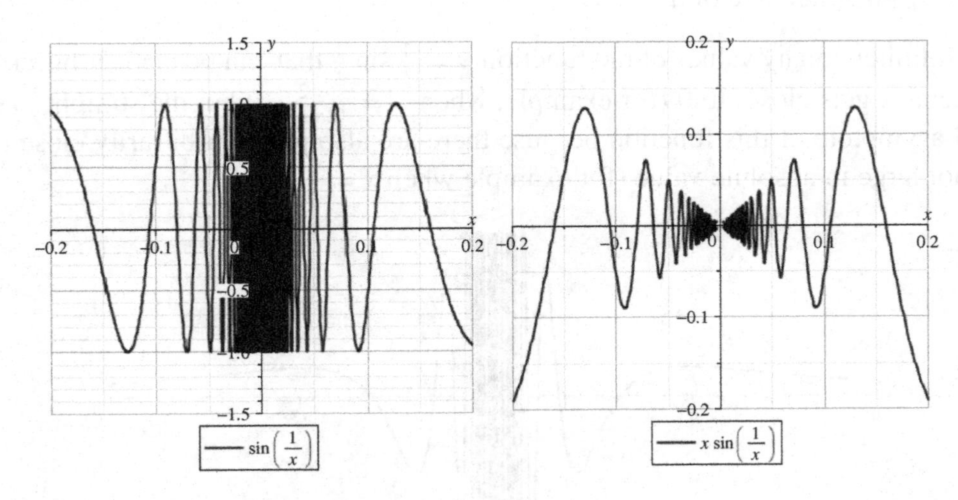

What functions could replace $f(x)$ in the counter-example?

Counter-Example 4.2.6

If a function $y = f(x)$ is not bounded in any neighbourhood of the point $x = a$ then either $\lim_{x\to a^+} |f(x)| = \infty$ or $\lim_{x\to a^-} |f(x)| = \infty$.

The function $f(x) = \frac{1}{x}\cos\frac{1}{x}$ is not bounded in any neighbourhood of the point $x = 0$ but neither $\lim_{x\to 0^+} |\frac{1}{x}\cos\frac{1}{x}|$ nor $\lim_{x\to 0^-} |\frac{1}{x}\cos\frac{1}{x}|$ exist.

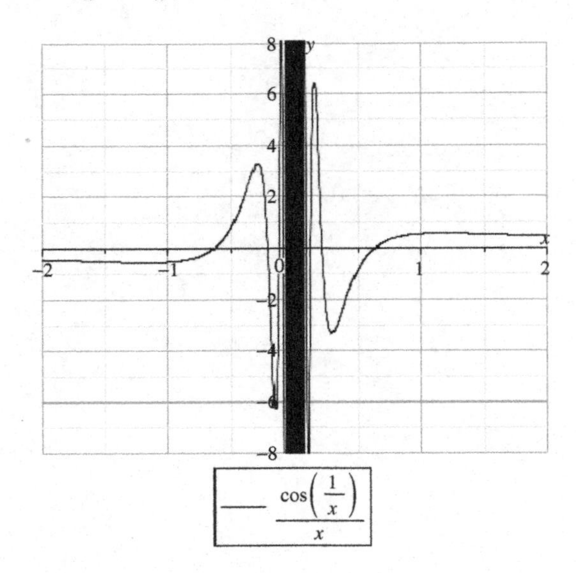

Why has the sine function of 2.5 been replaced by the cosine function for this counter-example?

Counter-Example 4.2.7

If a function $y = f(x)$ is continuous for all real x and $\lim_{n\to\infty} f(n) = A$ then $\lim_{x\to\infty} f(x) = A$.

For the continuous function $y = \cos(2\pi x)$ the limit $\lim_{n\to\infty} \cos(2\pi n)$ equals 1 because $\cos(2\pi n) = 1$ for any natural n but $\lim_{x\to\infty} \cos(2\pi x)$ does not exist.

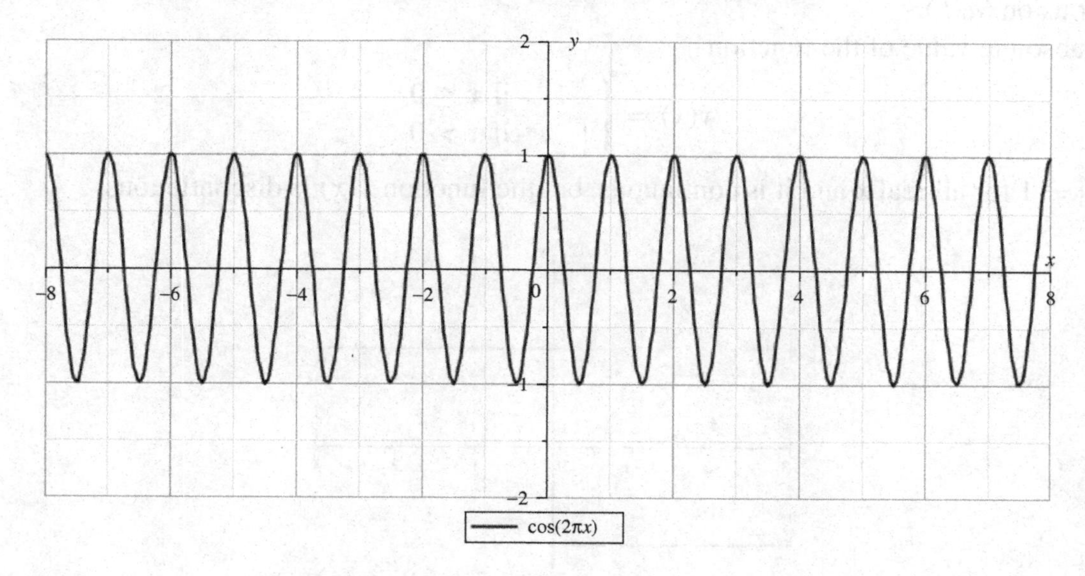

Comments. Statement 7 is the converse of the true statement:

$$\lim_{x\to\infty} f(x) = A \Rightarrow \lim_{n\to\infty} f(n) = A.$$

The example offered has a countable number of points (the integers) at which it takes the value A. Could a function take the value A on an even larger set, still meet the conditions, and yet be a counter-example? Modify the example so that $\lim_{x\to a} f(n) = A$ but f is not constant on n, and still gives a counter-example.

4.3. Continuity

Counter-Example 4.3.1

If the absolute value of the function $y = f(x)$ is continuous on (a, b) then the function is also continuous on (a, b).

The absolute value of the function

$$y(x) = \begin{cases} -1, & \text{if } x \leq 0 \\ 1, & \text{if } x > 0 \end{cases}$$

is $|y(x)| = 1$ for all real x and it is continuous, but the function $y(x)$ is discontinuous.

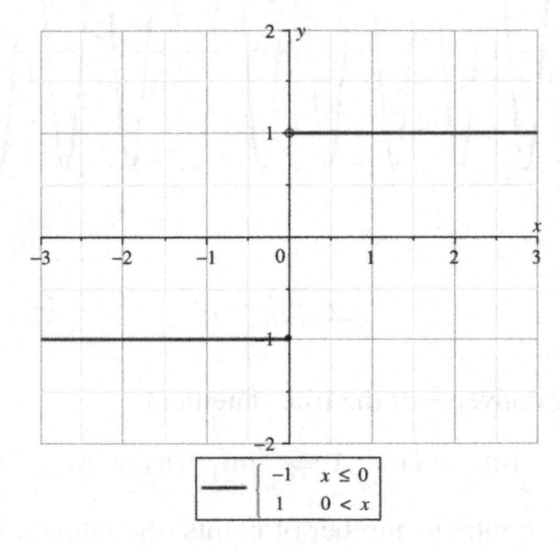

What is it about the example which makes it work as a counter-example? If $f(x)$ is continuous on an interval (a, b), must $|f(x)|$ be continuous? What about $f(|x|)$?

Counter-Example 4.3.2

If both functions $y = f(x)$ and $y = g(x)$ are discontinuous at $x = a$ then $f(x) + g(x)$ is also discontinuous at $x = a$.

$$f(x) = -\frac{1}{x-a}, \quad \text{if } x \neq a$$
$$g(x) = x + \frac{1}{x-a}, \quad \text{if } x \neq a$$
$$f(x) = g(x) = \frac{a}{2}, \quad \text{if } x = a.$$

Both functions $f(x)$ and $g(x)$ are discontinuous at $x = a$ but the function

$$f(x) + g(x) = \begin{cases} x, & \text{if } x \neq a \\ a, & \text{if } x = a \end{cases}$$

is continuous at $x = a$. For example, if $a = 2$:

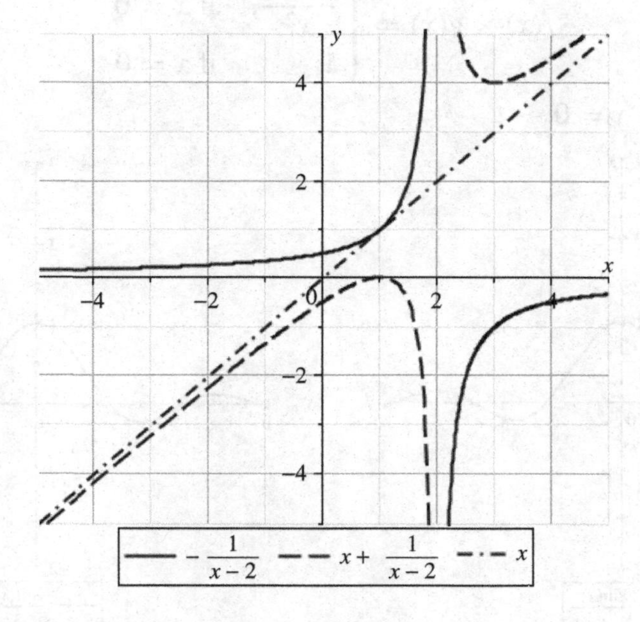

Let $h(x)$ be any continuous function on an interval (a, b), and $d(x)$ a function discontinuous at $x = c$ in that interval. Then $h(x) - d(x)$ is discontinuous at $x = c$ as is $h(x) + d(x)$, and their sum is continuous at $x = c$. How might this construction be generalised?

Counter-Example 4.3.3

If both functions $y = f(x)$ and $y = g(x)$ are discontinuous at $x = a$ then $f(x) \times g(x)$ is also discontinuous at $x = a$.

Both functions

$$f(x) = \begin{cases} \dfrac{\sin x}{x}, & \text{if } x \neq 0 \\ 2, & \text{if } x = 0 \end{cases}$$

and

$$g(x) = \begin{cases} \dfrac{\sin x}{x}, & \text{if } x \neq 0 \\ \dfrac{1}{2}, & \text{if } x = 0 \end{cases}$$

are discontinuous at the point $x = 0$ but their product

$$f(x) \times g(x) = \begin{cases} \dfrac{\sin^2 x}{x^2}, & \text{if } x \neq 0 \\ 1, & \text{if } x = 0 \end{cases}$$

is continuous at the point $x = 0$.

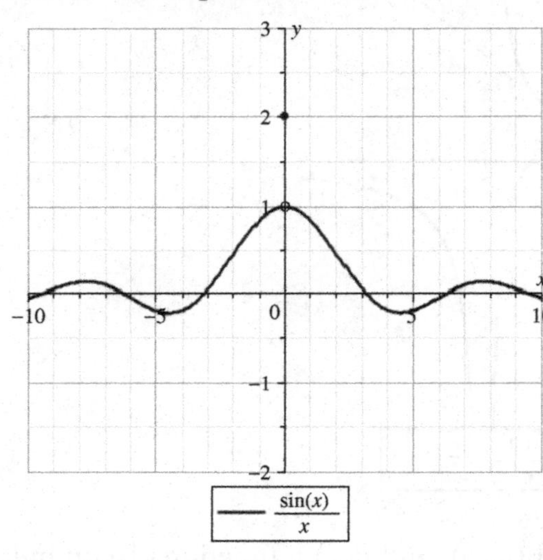

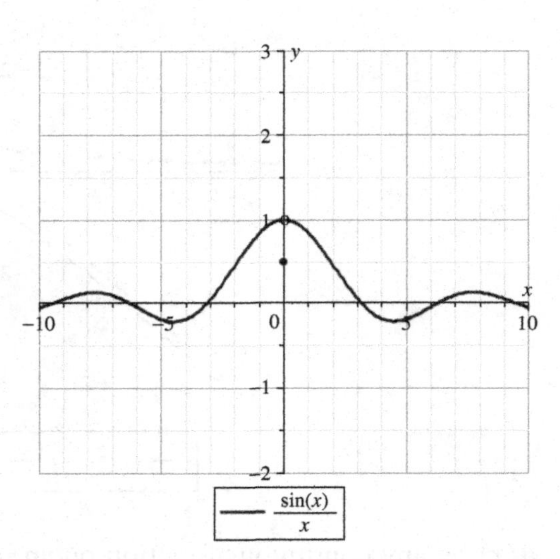

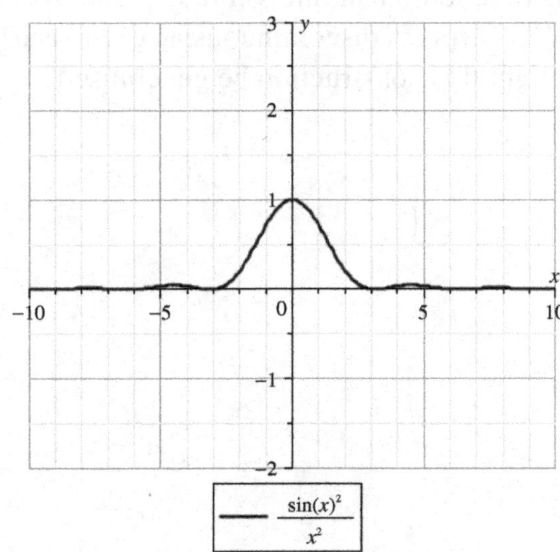

What is it about the example which makes it work as a counter-example? How might you generalise the example? For example, let $F(x)$ and $G(x)$ be any functions discontinuous at $x = a$ but continuous elsewhere.

Let

$$f(x) = \begin{cases} c & \text{if } x = a \\ F(x) & \text{elsewhere} \end{cases}$$

where $\lim_{x \to a} F(x) \neq c$ and let

$$g(x) = \begin{cases} \lim\limits_{x \to a} \dfrac{F(x)G(x)}{c} & \text{if } x = a \\ G(x) & \text{otherwise.} \end{cases}$$

Then $f(x)g(x)$ is also a counter-example to the conjecture.

Counter-Example 4.3.4

A function always has a local maximum between any two local minima.

The functions $y = \frac{x^4+0.1}{x^2}$ and $y = \sec^2 x$ have no maximum between two local minima:

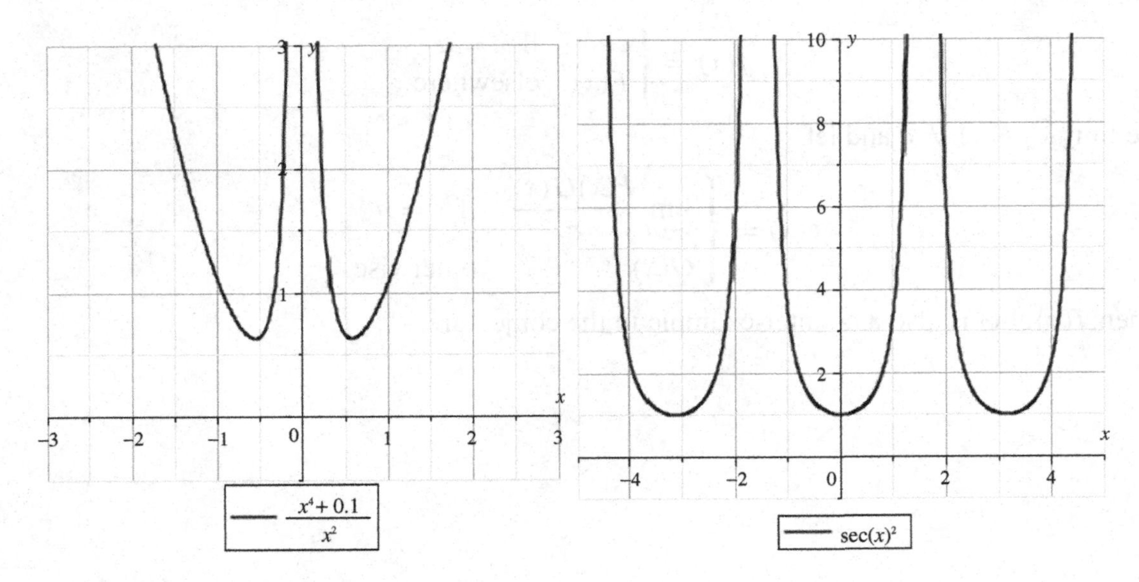

What if the function is also continuous? What if it is bounded on the interval? Is there any other form of discontinuity which would work as a counter-example? For instance, does

$$f(x) = \begin{cases} \sin\left(\frac{1}{x}\right) & \text{if } x \neq 0 \\ 0 & \text{if } x = 0 \end{cases}$$

work as a counter-example?

There are functions which have a local maximum at every rational: Consider

$$f(x) = \begin{cases} 0 & \text{if } x \text{ is irrational} \\ \frac{1}{q} & \text{if } x \text{ is } \frac{p}{q} \text{ in lowest terms.} \end{cases}$$

If you are at $x = p/q$, then any rational within $1/10q$, say, must have smaller value, and at irrationals the value is 0. Is this a counter-example to the conjecture?

Counter-Example 4.3.5

For a continuous function there is always a local maximum between any two local minima.

The continuous function below does not have a local maximum between its two local minima.

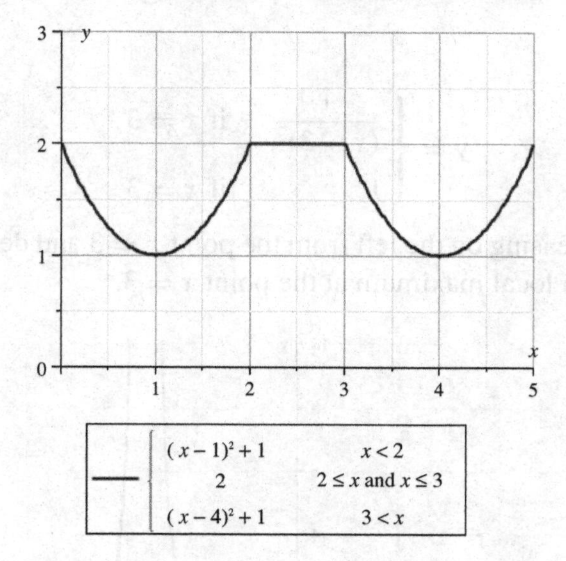

Comments. A *strict* inequality in the definition of a local maximum is accepted here: A function $y = f(x)$ has a local maximum at the point $x = a$ if $f(a) > f(x)$ for all x within a certain neighbourhood $(a - \delta, a + \delta)$, $\delta > 0$ of the point $x = a$. Otherwise in the above graph we have to treat each point of the line segment as a local maximum.

Counter-Example 4.3.6

If a function is defined in a certain neighbourhood of point $x = a$ including the point itself and is increasing on the left from $x = a$ and decreasing on the right from $x = a$, then there is a local maximum at $x = a$.

The function

$$y = \begin{cases} \dfrac{1}{(x-3)^2}, & \text{if } x \neq 3 \\ 1, & \text{if } x = 3 \end{cases}$$

is defined for all real x, increasing on the left from the point $x = 3$ and decreasing on the right from the point $x = 3$ but has no a local maximum at the point $x = 3$.

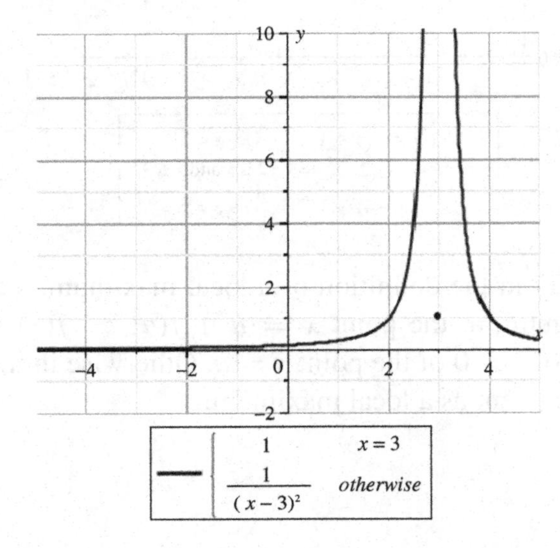

What could be changed in the example and still it is a counter-example? What has to be done to convert local maxima to local minima?

Counter-Example 4.3.7

If a function is defined on $[a, b]$ and continuous on (a, b) then it takes its extreme values on $[a, b]$.

The function

$$y = \begin{cases} \tan x, & \text{if } x \in \left(-\frac{\pi}{2}, \frac{\pi}{2}\right) \\ 0, & \text{if } x = \pm\frac{\pi}{2} \end{cases}$$

is defined on $[-\frac{\pi}{2}, \frac{\pi}{2}]$ and is continuous on $(-\frac{\pi}{2}, \frac{\pi}{2})$ but it has no extreme values on $[-\frac{\pi}{2}, \frac{\pi}{2}]$.

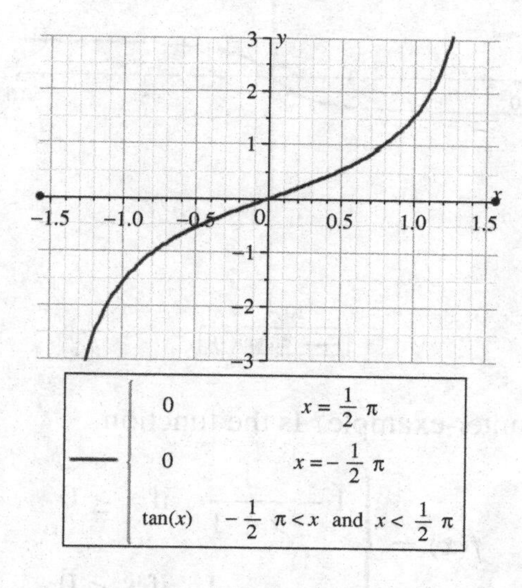

Use the function $f(x) = 1/x$ to construct a similar counter-example.

Counter-Example 4.3.8

Every continuous and bounded function on $(-\infty, \infty)$ takes on its extreme values.

The function $f(x) = \tan^{-1}(x)$ is continuous and bounded on $(-\infty, \infty)$ but takes no extreme values.

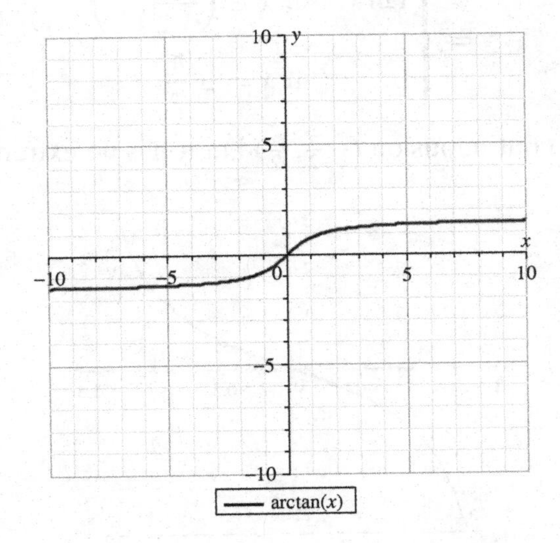

Why is $f(x) = \sqrt[3]{x}$ not a counter-example? Is the function

$$f(x) = \begin{cases} 1 - \dfrac{1}{x+1} & \text{if } x \geq 0 \\ \dfrac{1}{1-x} - 1 & \text{if } x < 0 \end{cases}$$

a counter-example? What distinguishes it from $\tan^{-1}x$ as a counter-example?

Counter-Example 4.3.9

If a function $y = f(x)$ is continuous on $[a, b]$, the tangent line exists at all points on its graph and $f(a) = f(b)$ then there is a point c in (a, b) such that the tangent line at the point $(c, f(c))$ is horizontal.

The function $y = f(x)$ below is continuous on $[-1, 1]$, the tangent line exists at all points on the graph and $f(-1) = f(1)$ but there is no point c in $(-1, 1)$ such that the tangent line at the point $(c, f(c))$ is horizontal.

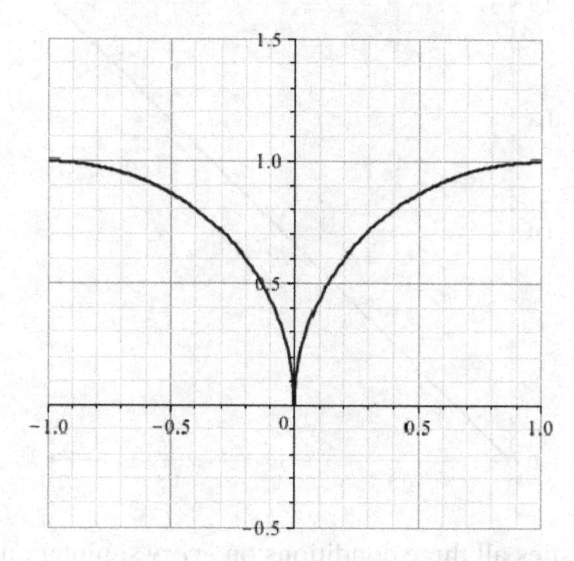

Must there be a point at which the tangent line is either horizontal or vertical in order to be a counter-example? In other words, are there other types of counter-examples?

Counter-Example 4.3.10

If on the closed interval $[a, b]$ a function is bounded, takes its maximum and minimum values and takes all its values between the maximum and minimum values then this function is continuous on $[a, b]$.

The function $f(x) = x$ for $0 < x < 2$, but $f(0) = 2$, $f(2) = 0$ satisfies the three conditions above, but is not continuous on $[0, 2]$.

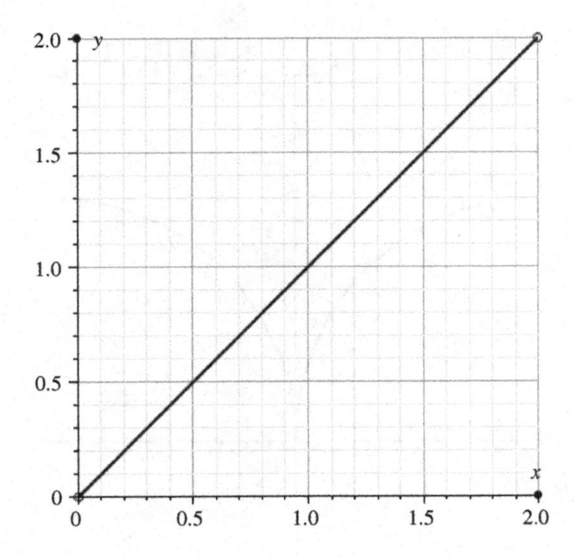

What if the function satisfies all three conditions on *every* subinterval?

Counter-Example 4.3.11

If on the closed interval $[a, b]$ a function is bounded, takes its maximum and minimum values and takes all its values between the maximum and minimum values then this function is continuous at some points or subintervals on $[a, b]$.

The function

$$f(x) = \begin{cases} 1, & \text{if } x = 0 \\ x, & \text{if } x \text{ is rational}, \quad x \neq 1 \\ -x, & \text{if } x \text{ is irrational}, \quad x \neq 1, x \neq -1 \\ 0, & \text{if } x = 1 \end{cases}$$

satisfies all three conditions above but it is discontinuous at *every* point on $[-1, 1]$. It is impossible to draw the graph of the function $y = f(x)$ but the sketch below gives an idea of its behaviour. Again, fine dots indicate that some but not all points on the implied line are present.

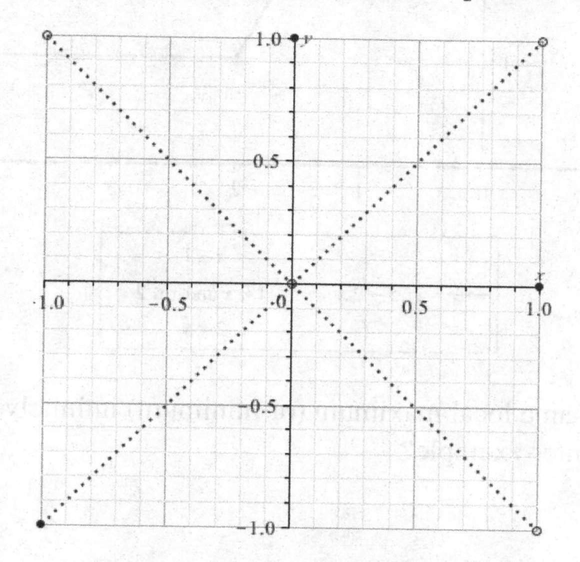

What could replace rational/irrational as a distinction between the specified values? Why is the condition imposed that $f(x) = 0$ if $x = 1$ in the counter-example?

Counter-Example 4.3.12

If a function is continuous on $[a, b]$ then it cannot take its absolute maximum or minimum value infinitely many times.

The function below takes its absolute maximum value (3) and its absolute minimum value (1) an infinite number of times on the interval $[-1, 4]$.

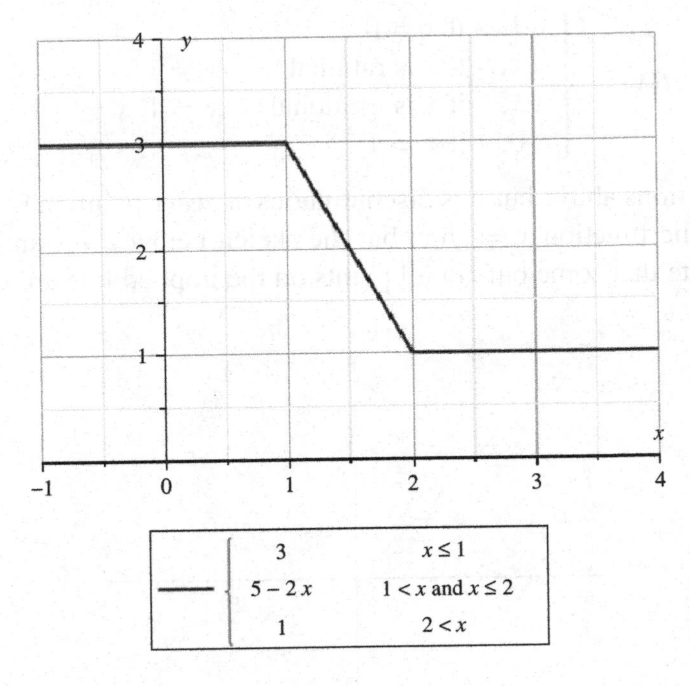

$$\begin{cases} 3 & x \le 1 \\ 5 - 2x & 1 < x \text{ and } x \le 2 \\ 1 & 2 < x \end{cases}$$

What about having the same local maximum (or minimum) infinitely many times? Would $x \sin^2 x$ be a counter-example?

Counter-Example 4.3.13

If a function $y = f(x)$ is defined on $[a, b]$ and $f(a) \times f(b) < 0$ then there is some point $c \in (a, b)$ such that $f(c) = 0$.

The function

$$f(x) = \begin{cases} \dfrac{1}{x}, & \text{if } x \neq 0 \\ 1, & \text{if } x = 0 \end{cases}$$

is defined on $[-1, 1]$ and $f(-1) \times f(1) = (-1) \times (1) = -1 < 0$ but there is no point c on $[-1, 1]$ such that $f(c) = 0$.

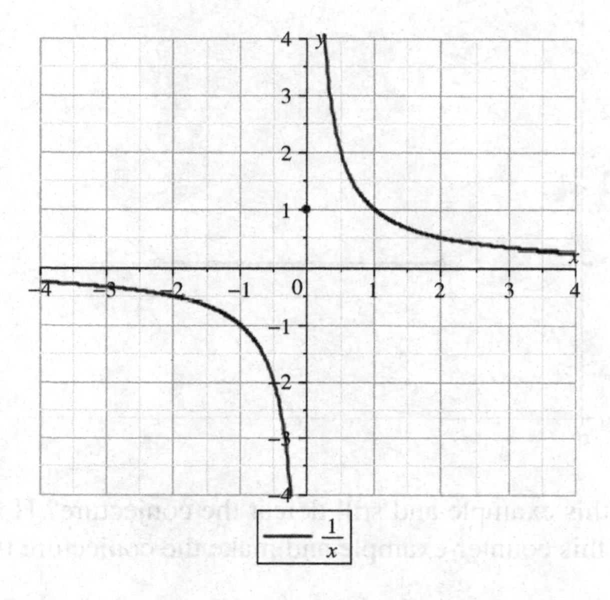

What is the key to making this counter-example work? How does that inform other choices of counter-example?

Counter-Example 4.3.14

If a function $y = f(x)$ is defined on $[a, b]$ and continuous on (a, b) then for any $N \in (f(a), f(b))$ there is some point $c \in (a, b)$ such that $f(c) = N$.

The function below is defined on $[1, 3]$ and continuous on $(1, 3)$ but for no N between $f(1)$ and $f(3)$ is there a corresponding c for which $f(c) = N$. In other words, for any $N \in (f(1), f(3))$ there is no point $c \in (1, 3)$ such that $f(c) = N$.

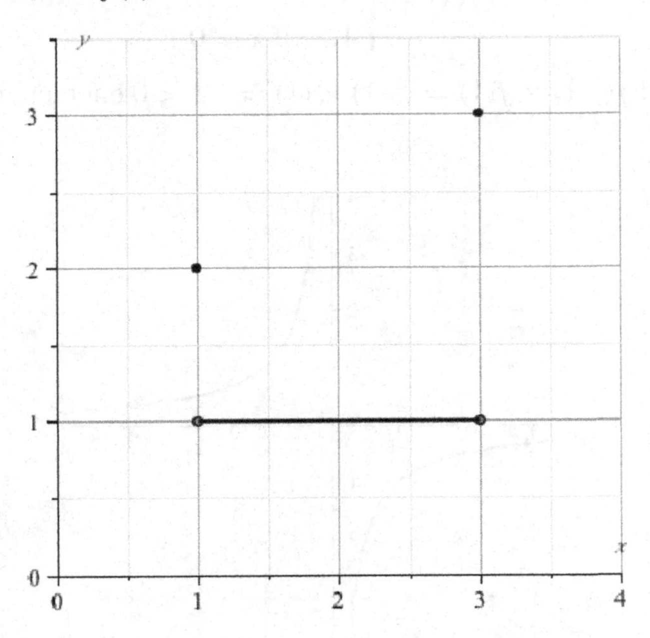

What can you vary in this example and still defeat the conjecture? How could you modify the conjecture so as to defeat this counter-example and make the conjecture true?

Counter-Example 4.3.15

If a function is discontinuous at every point in its domain then the square and the absolute value of this function cannot be continuous.

The function

$$f(x) = \begin{cases} 1, & \text{if } x \text{ is rational} \\ -1, & \text{if } x \text{ is irrational} \end{cases}$$

is discontinuous at every point in its domain but both the square and the absolute value $f^2(x) = |f(x)| = 1$ are continuous. It is impossible to draw the graph of the function $y = f(x)$ but the sketch below gives an idea of its behaviour.

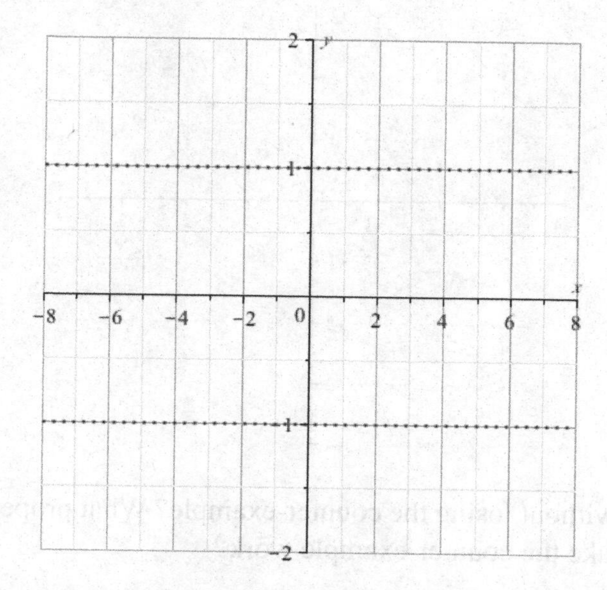

What can be changed and still preserve the counter-example? For example, rationals could be replaced by any dense subset of R with dense complement.

Counter-Example 4.3.16

A function cannot be continuous at only one point in its domain and discontinuous everywhere else.

The function

$$g(x) = \begin{cases} x, & \text{if } x \text{ is rational} \\ -x, & \text{if } x \text{ is irrational} \end{cases}$$

is continuous at the point $x = 0$ and discontinuous at all other points on R. It is impossible to draw the graph of the function $y = g(x)$ but the sketch below gives an idea of its behaviour.

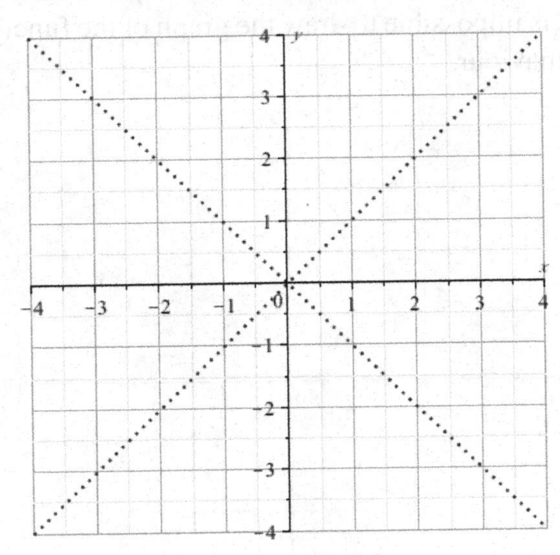

What can be changed without losing the counter-example? What properties do the rationals and irrationals share which make the counter-example work?

Counter-Example 4.3.17

A sequence of continuous functions on $[a, b]$ always converges to a continuous function on $[a, b]$.

The sequence of continuous functions $f_n(x) = x^n$, $n \in N$ on $[0, 1]$ converges to a discontinuous function when $n \to \infty$:

$$\lim_{n\to\infty} f_n(x) = \begin{cases} 0, & \text{if } x \in [0, 1) \\ 1, & \text{if } x = 1. \end{cases}$$

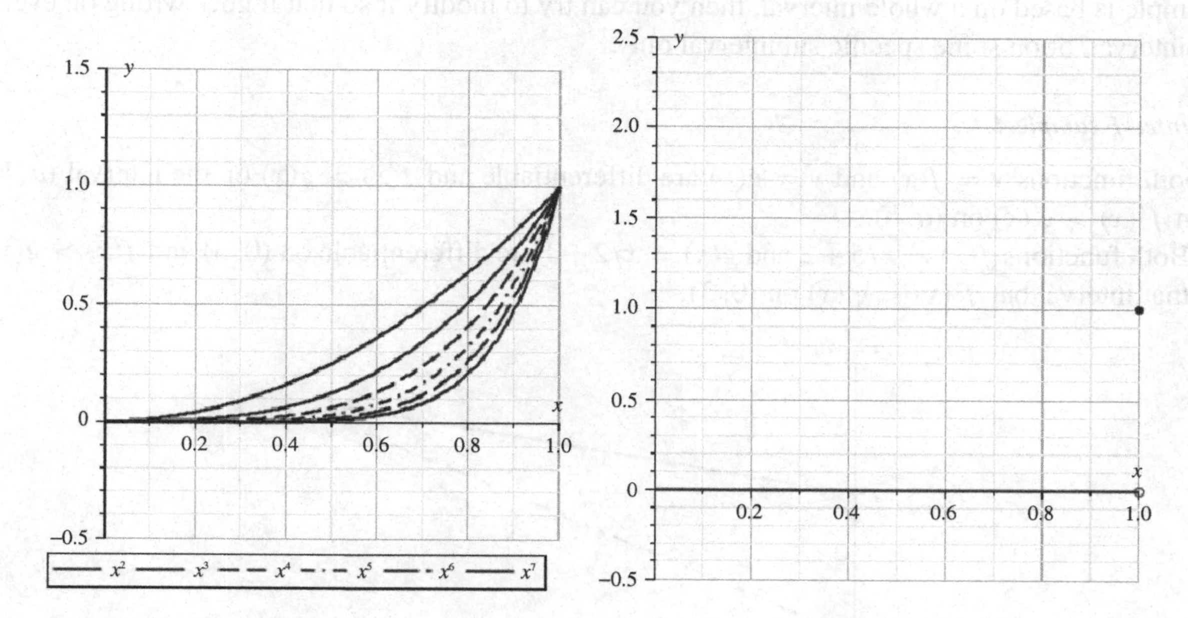

What about $f(x) = x^{\frac{1}{n}}$ for various n?

4.4. Differential Calculus

As differentiation is usually presented as a phenomenon occurring at a point (a function is differentiable at a point c, in the interval (a, b), perhaps even at every such point in the interval) there are general strategies for constructing and then modifying counter-examples. If a counter-example is based on something going wrong at just one point, then you can try to modify it so that it goes wrong at finitely many points, then infinitely many, then at every point in an interval. If a counter-example is based on a whole interval, then you can try to modify it so that it goes wrong on every subinterval, or on some specific subinterval only.

Counter-Example 4.4.1

If both functions $y = f(x)$ and $y = g(x)$ are differentiable and $f(x) > g(x)$ on the interval (a, b) then $f'(x) > g'(x)$ on (a, b).

Both functions $f(x) = x/5 + 2$ and $g(x) = x/2 + 1$ are differentiable on $(0, 3)$ and $f(x) > g(x)$ on that interval but $f'(x) < g'(x)$ on $(0, 3)$.

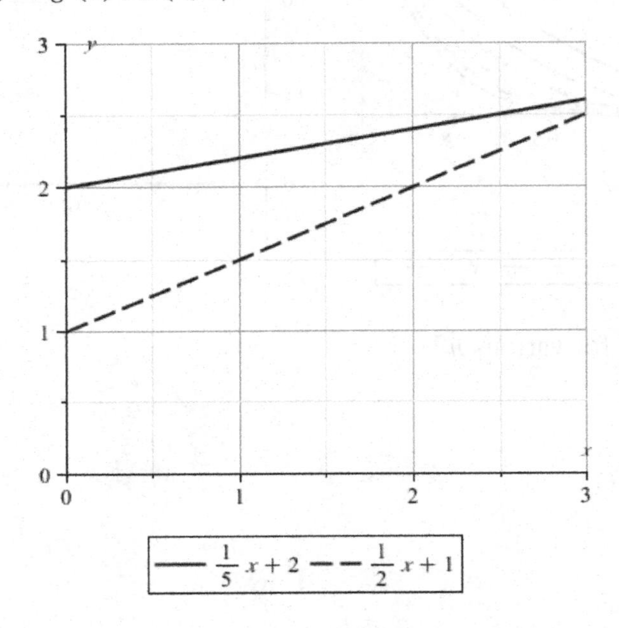

In order to appreciate the generality of which this counter-example is only a single instance, construct examples which are counter-examples on subintervals only, or at a single point only. Given $f(x)$ differentiable on (a, b) construct families of $g(x)$ for which $f(x) > g(x)$ but $f'(x) < g'(x)$ on (a, b) or on some specified subinterval.

What about a converse conjecture: If $f'(x) > g'(x)$ on (a, b), then $f(x) > g(x)$ on that interval. The counter-examples to 4.4.1 can be modified to be counter-examples to this as well. However, if the interval is the whole of R, is there a counter-example valid for every point x, or must there always be some interval on which $f(x) > g(x)$?

Counter-Example 4.4.2

If a non-linear function is differentiable and monotone on $(0, \infty)$ then its derivative is also monotone on $(0, \infty)$.

The non-linear function $y = x + \sin x$ is differentiable and monotone on $(0, \infty)$ but its derivative $y' = 1 + \cos x$ is not monotone on $(0, \infty)$.

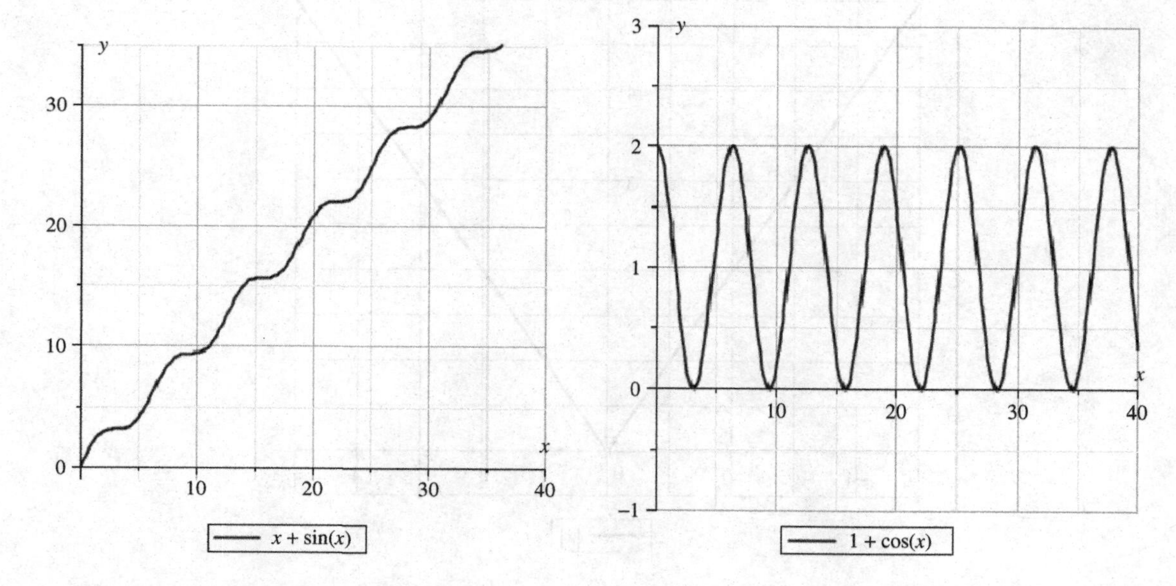

Tinker with the example given to try to find the essential idea which makes the example into a counter-example. What conditions must a differentiable function $g(x)$ satisfy if $g(x)$ is not monotone, but $g(x) + ax$ is monotone for some constant a?

Counter-Example 4.4.3

If a function is continuous at a point then it is differentiable at that point:

The function $y = |x|$ is continuous at the point $x = 0$ but it is not differentiable at that point.

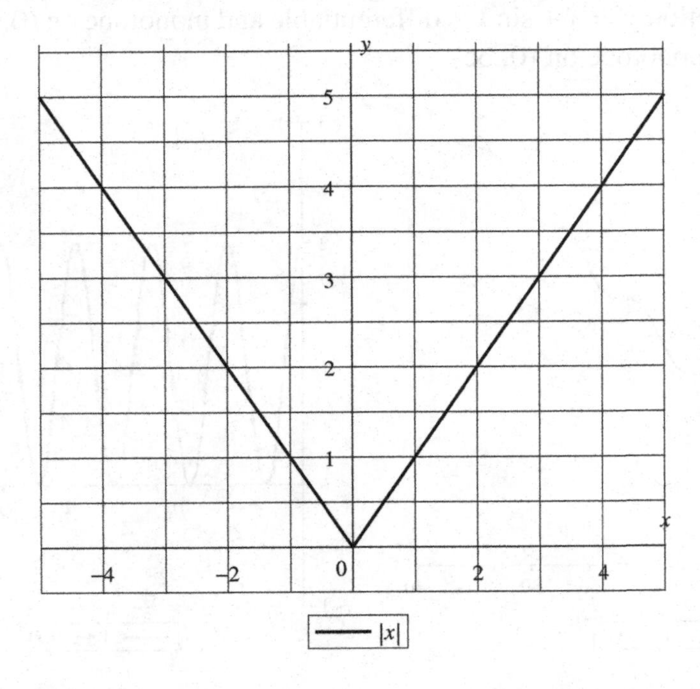

Use $|x|$ to construct continuous functions which are not differentiable at n points (for $n = 2, 3, \ldots$), and at countably many points.

What changes could be made to $|x|$ and still provide a counter-example? For example what about the function

$$f(x) = \begin{cases} x^{\frac{1}{2}} & \text{if } x \geq 0 \\ -(-x)^{\frac{1}{2}} & \text{if } x < 0 \end{cases}?$$

Counter-Example 4.4.4

If a function is continuous on R and the tangent line exists at any point on its graph then the function is differentiable at any point on R.

The function $y = \sqrt[3]{x^2}$ is continuous on R and the tangent line exists at any point on its graph but the function is not differentiable at the point $x = 0$.

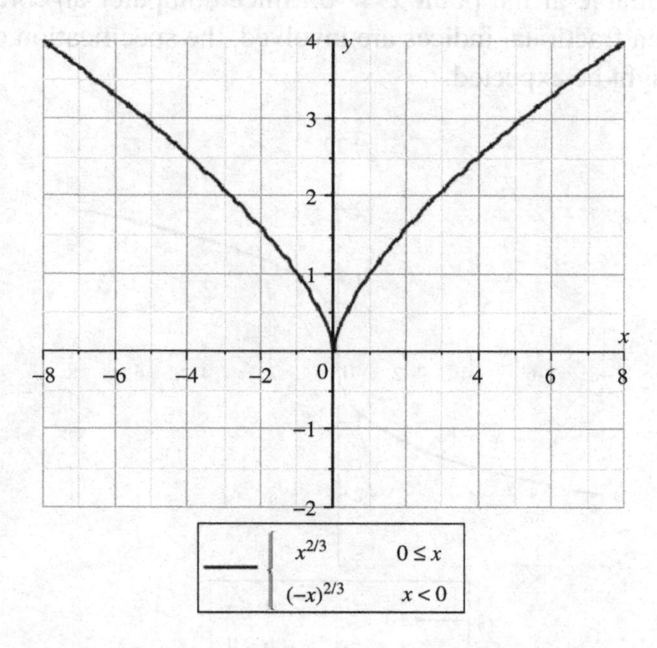

What can you change in $y = x^{\frac{2}{3}}$ and still have a counter-example? Do you need to have the same function on both sides of the origin? What is special about the origin?

Counter-Example 4.4.5

If a function is continuous on the interval (a, b) and its graph is a *smooth* curve (no sharp corners) on that interval then the function is differentiable at any point on (a, b).

(a) The function $y = \sqrt[3]{x}$ is continuous on R and its graph is a smooth curve (no sharp corners), but it is not differentiable at the point $x = 0$. Since computer algebra systems require x to be non-negative when fractional indices are involved, the specification of the function is more complicated than might be expected.

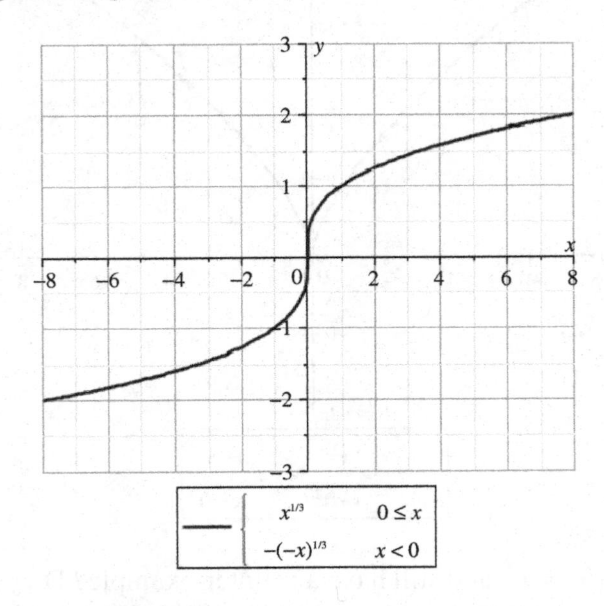

(b) The function below is continuous on R and its graph is a smooth curve (no sharp corners), but it is non-differentiable at infinitely many points on R.

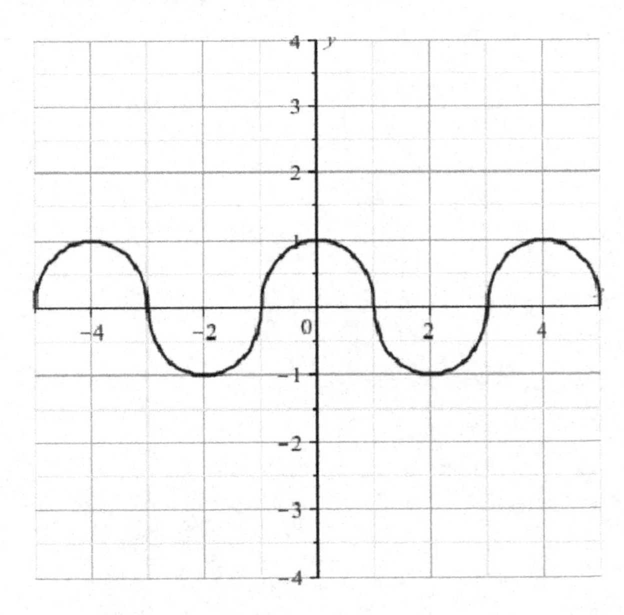

Constructing functions which are non-differentiable at infinitely many points on a finite interval such as [0, 1] can be achieved by scaling and translating the individual components.

Counter-Example 4.4.6

If the derivative of a function is zero at a point then the function is neither increasing nor decreasing at this point.

The derivative of the function $y = x^3$ is zero at the point $x = 0$ but the function is increasing at this point.

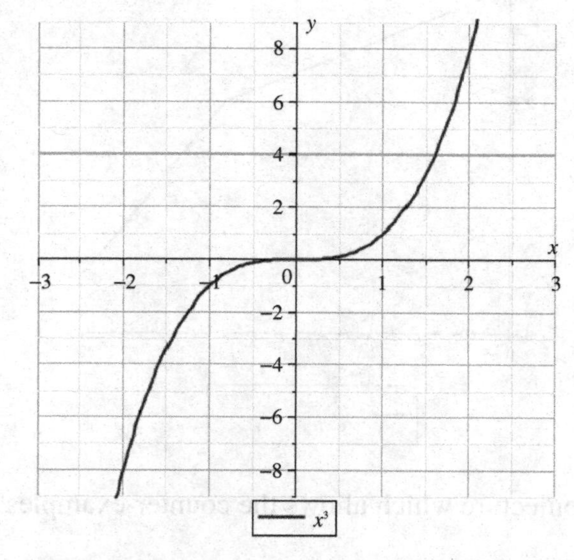

Increasing at x means that for some $\delta > 0$, $f(x+t) > f(x)$ for all $t \in (0, \delta)$ and $f(x-t) < f(x)$ for all $t \in (0, \delta)$.

Counter-Example 4.4.7

If a function is differentiable and decreasing on (a, b) then its gradient is negative on (a, b).

The function $y = -x^3$ is differentiable and decreasing on R but its gradient is zero at the point $x = 0$.

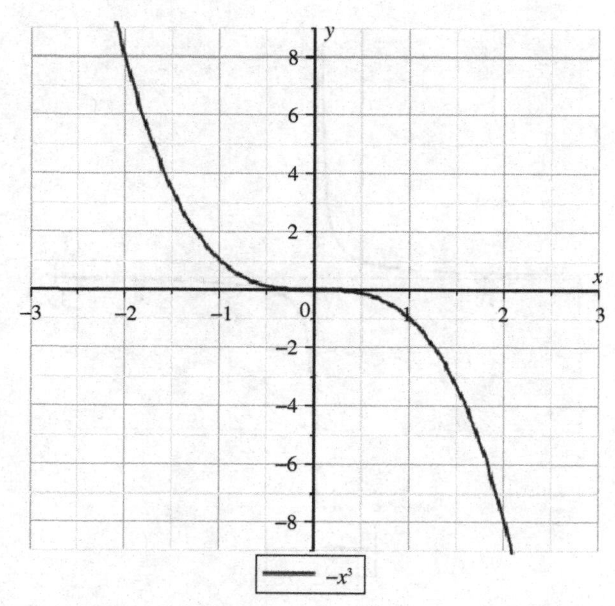

Modify the conjecture (and prove it) in order to avoid this type of counter-example. Is it just that $f'(x) = 0$ for some x which makes this a counter-example?

Counter-Example 4.4.8

If a function is continuous and decreasing on (a, b) then its gradient is non-positive on (a, b).

The function below is continuous and decreasing on R but its gradient does not exist at the point $x = 1$.

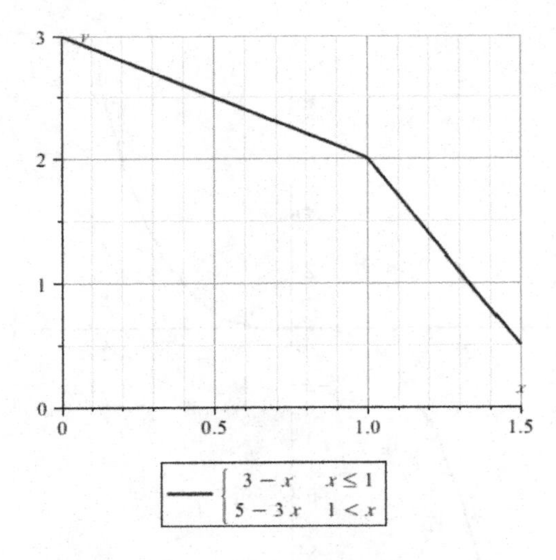

What is missing in the conjecture which allows the counter-examples?

Counter-Example 4.4.9

If a function has a positive derivative at every point in its domain then the function is increasing everywhere in its domain.

The derivative of the function $y = -\frac{1}{x}(x \neq 0)$ is $y' = \frac{1}{x^2}$, which is positive for all $x \neq 0$.

According to the definition, a function is increasing in its domain if for any x_1, x_2 from its domain from $x_1 < x_2$ it follows that $f(x_1) < f(x_2)$. If we take $x_1 = -1$ and $x_2 = 1(x_1 < x_2)$ then in this case $f(x_1) > f(x_2)$.

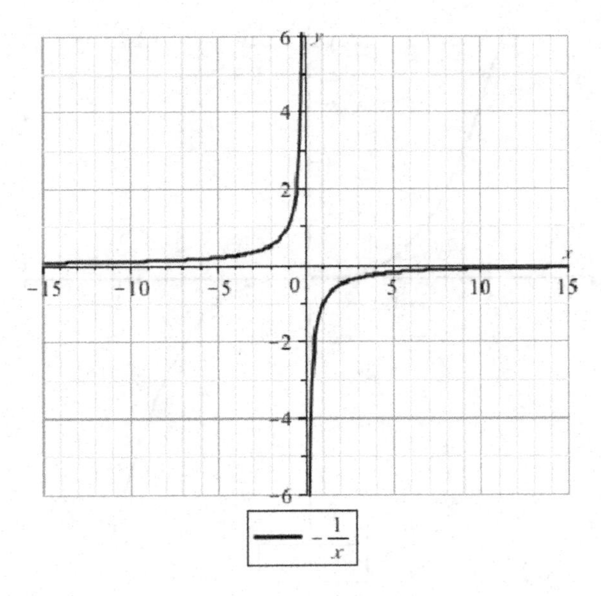

What if the function is specified on an interval, so there are no gaps in its domain?

Counter-Example 4.4.10

If a function $y = f(x)$ is defined on $[a, b]$ and has a local maximum at the point $c \in (a, b)$ then in a sufficiently small neighbourhood of the point $x = c$ the function is increasing on the left and decreasing on the right from $x = c$.

The function below is defined on [0, 2] and has a maximum at the point 1 in [0, 2] but it is neither increasing on the left nor decreasing on the right from the point $x = 1$.

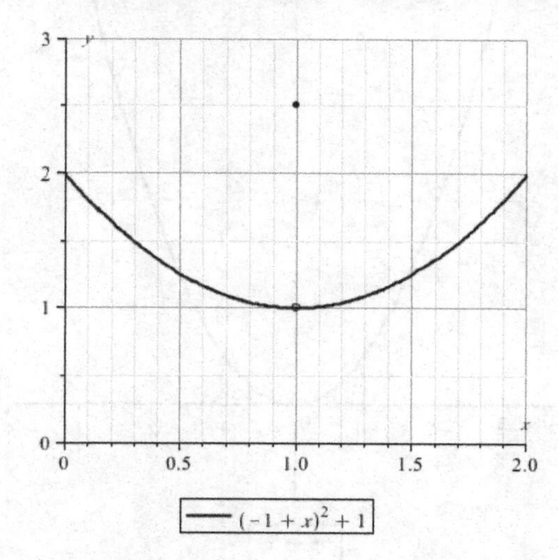

Comments. The definition of a local maximum requires neither differentiability nor continuity of a function at the point of interest: A function $y = f(x)$ has a local maximum at the point $x = c$ if $f(c) > f(x)$ for all x within a certain neighbourhood $(c - \delta, c + \delta)$, $\delta > 0$ of the point $x = c$.

What additional assumptions about f are needed to make the conjecture valid?

Counter-Example 4.4.11

If a function $y = f(x)$ is differentiable for all real x and $f(0) = f'(0) = 0$ then $f(x) = 0$ for all real x.

Both the function $y = x^2$ and its derivative $y' = 2x$ equal zero at the point $x = 0$ but the function is not zero for all real x.

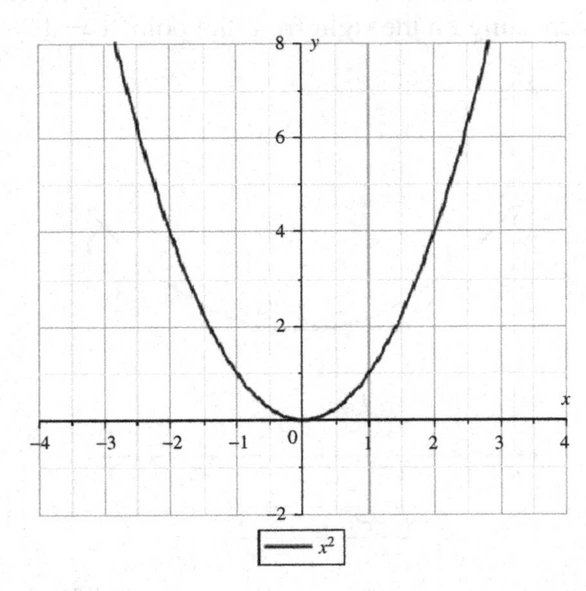

There are functions for which $f^{(n)}(0) = 0$ for all positive integers n, and yet $f(x) \neq 0$. For example, $f(x) = e^{\frac{-1}{x^2}}$ $(x \neq 0)$ and $f(0) = 0$ has all its derivatives at 0 being 0, without itself being identically 0.

Counter-Example 4.4.12

If a function $y = f(x)$ is differentiable on the interval (a, b) and takes both positive and negative values on (a, b) then its absolute value $|f(x)|$ is not differentiable at the point(s) where $f(x) = 0$, e.g. $|f(x)| = |x|$ or $|f(x)| = |\sin x|$.

The function $y = x^3$ is differentiable on R and takes both positive and negative values but its absolute value $y = |x^3|$ is differentiable at the point $x = 0$ where the function equals zero.

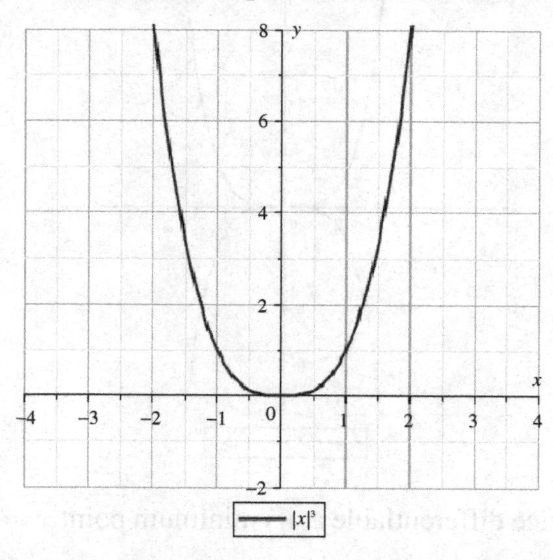

Comments. To make the statement true it should conclude: "…then its absolute value $|f(x)|$ is not differentiable at the points where both $f(x) = 0$ and $f'(x) \neq 0$."

Counter-Example 4.4.13

If both functions $y = f(x)$ and $y = g(x)$ are differentiable on the interval (a, b) and intersect somewhere on (a, b) then the function $\max\{f(x), g(x)\}$ is not differentiable at the point(s) where $f(x) = g(x)$.

The function $\max\{x^3, x^4\}$ on $(-1, 1)$ is differentiable at the point $x = 0$ where the functions $y = x^3$ and $y = x^4$ intersect.

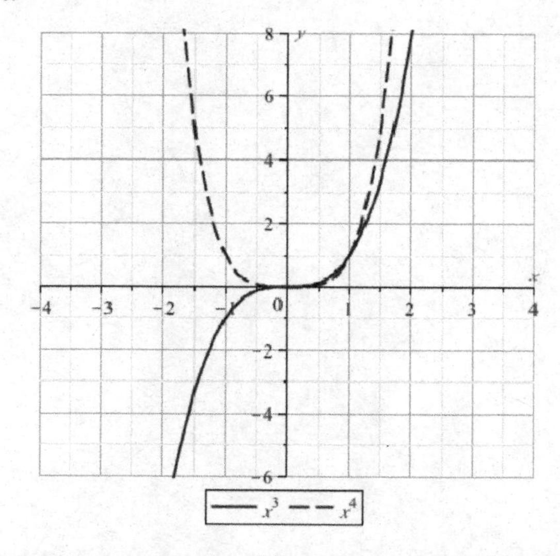

Comments. To make the statement true it should conclude: "…then the function $\max\{f(x), g(x)\}$ is not differentiable at the point(s) where both $f(x) = g(x)$ and $f'(x) \neq g'(x)$."

Counter-Example 4.4.14

If a function is twice differentiable at a local maximum (minimum) point then its second derivative is negative (positive) at that point.

The function $y = -x^4$ is twice differentiable at its maximum point $x = 0$ but the second derivative is zero at this point.

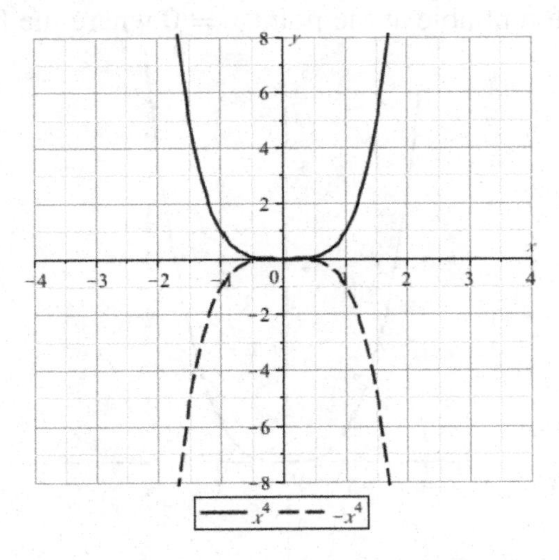

The function $y = x^4$ is twice differentiable at its minimum point $x = 0$ but the second derivative is zero at that point.

What do these counter-examples say about seeking the maximum and minimum of differentiable functions?

Counter-Example 4.4.15

If both functions $y = f(x)$ and $y = g(x)$ are non-differentiable at $x = a$ then $f(x) + g(x)$ is also not differentiable at $x = a$.

Both functions $f(x) = |x|$ and $g(x) = -|x|+1$ are not differentiable at $x = 0$ but $f(x)+g(x) = 1$ is differentiable at every x including $x = 0$.

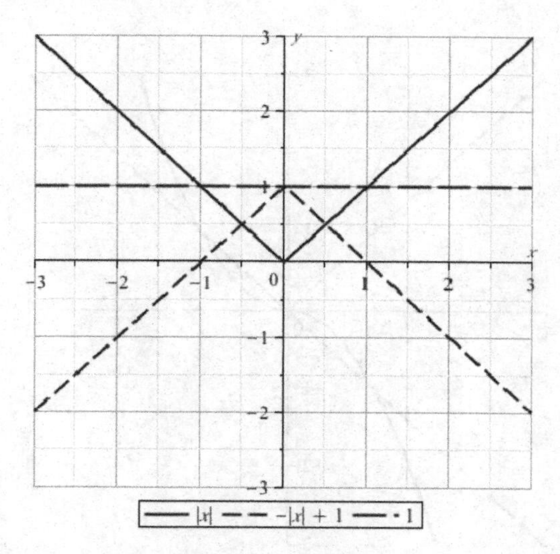

<u>Comments.</u> More generally, $f(x) = A(x)$ and $g(x) = B(x) - A(x)$, where $A(x)$ is not differentiable and $B(x)$ is differentiable at $x = a$. Both $f(x)$ and $g(x)$ are not differentiable, but $f(x)+g(x) = B(x)$ is differentiable at $x = a$.

Start with any differentiable function and represent it as the sum of two functions which are not differentiable at a single point; at a finite number of points; at infinitely many points; at any points.

Counter-Example 4.4.16

If a function $y = f(x)$ is differentiable and a function $y = g(x)$ is not differentiable at $x = a$ then $f(x) \times g(x)$ is not differentiable at $x = a$.

The function $f(x) = x$ is differentiable at $x = 0$ and the function $g(x) = |x|$ is not differentiable at $x = 0$, but the function $f(x) \times g(x) = x|x|$ is differentiable at the point $x = 0$ (the derivative equals zero).

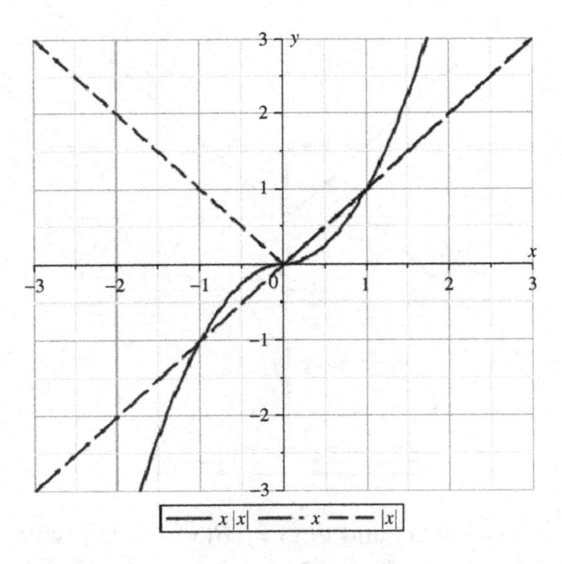

What is the same and what is different about this conjecture and the previous one, and about their counter-examples?

Counter-Example 4.4.17

If both functions $y = f(x)$ and $y = g(x)$ are not differentiable at $x = a$ then $f(x) \times g(x)$ is also not differentiable at $x = a$.

Both functions $f(x) = |x|$ and $g(x) = -|x|$ are not differentiable at the point $x = 0$ but the function $f(x) \times g(x) = -|x|^2 = -x^2$ is differentiable at $x = 0$.

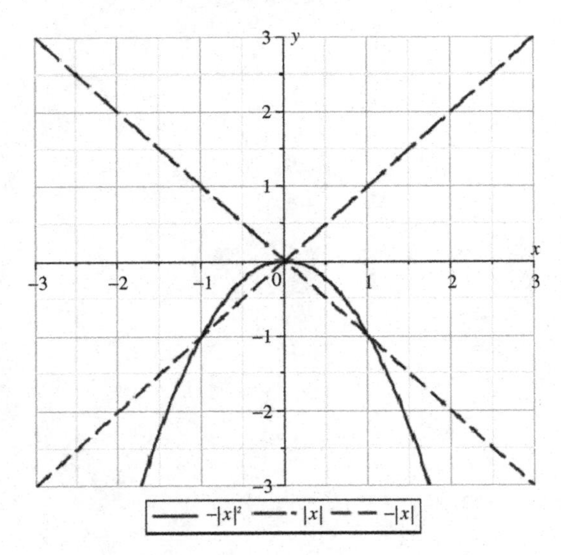

What is the same and what is different between 4.15 and 4.17?

Counter-Example 4.4.18

If a function $y = g(x)$ is differentiable at $x = a$ and a function $y = f(x)$ is not differentiable at $g(a)$ then the function $F(x) = f(g(x))$ is not differentiable at $x = a$.

The function $g(x) = x^2$ is differentiable at $x = 0$, and the function $f(x) = |x|$ is not differentiable at $g(0) = 0$, but the function $F(x) = f(g(x)) = |x^2| = x^2$ is differentiable at $x = 0$.

What makes this counter-example work? How could it be altered and still be a counter-example?

Counter-Example 4.4.19

If a function $y = g(x)$ is not differentiable at $x = a$ and a function $y = f(x)$ is differentiable at $g(a)$ then the function $F(x) = f(g(x))$ is not differentiable at $x = a$.

The function $g(x) = |x|$ is not differentiable at $x = 0$, the function $f(x) = x^2$ is differentiable at $g(0) = 0$, but the function $F(x) = f(g(x)) = |x|^2 = x^2$ is differentiable at $x = 0$.

Note that if g is differentiable at a point $x = a$, and f is differentiable at $g(a)$ then $f(g(x))$ is also differentiable at $x = a$.

Counter-Example 4.4.20

If a function $y = g(x)$ is not differentiable at $x = a$ and a function $y = f(x)$ is not differentiable at $g(a)$ then the function $F(x) = f(g(x))$ is not differentiable at $x = a$.

The function $g(x) = \frac{2}{3}x - \frac{1}{3}|x|$ is not differentiable at $x = 0$ and the function $f(x) = 2x + |x|$ is not differentiable at $g(0) = 0$, but the function $F(x) = f(g(x)) = 2\left(\frac{2}{3}x - \frac{1}{3}|x|\right) + \left|\frac{2}{3}x - \frac{1}{3}|x|\right|$ is differentiable at $x = 0$.

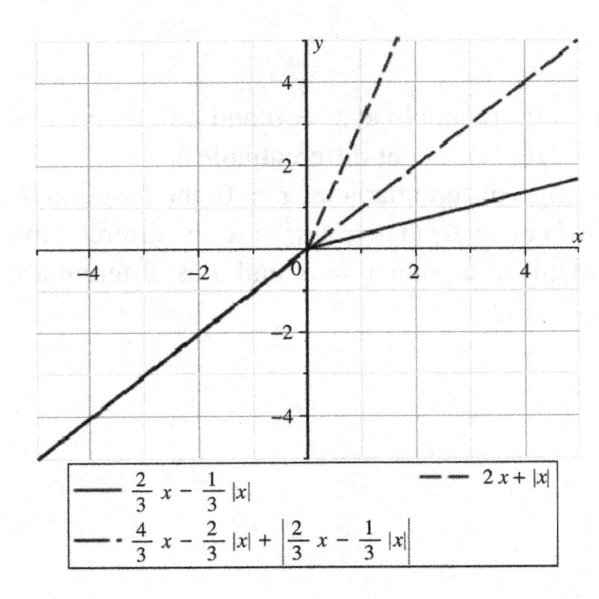

It is instructive to show this using the definition of the derivative.

$$F'(0) = \lim_{\Delta x \to 0} \frac{F(\Delta x) - F(0)}{\Delta x}$$

$$= \lim_{\Delta x \to 0} \frac{2\left(\frac{2}{3}\Delta x - \frac{1}{3}|\Delta x|\right) + \left|\frac{2}{3}\Delta x - \frac{1}{3}|\Delta x|\right|}{\Delta x}.$$

If $\Delta x \to 0^-$ then

$$\lim_{\Delta x \to 0^-} \frac{2\left(\frac{2}{3}\Delta x + \frac{1}{3}\Delta x\right) + \left|\frac{2}{3}\Delta x + \frac{1}{3}\Delta x\right|}{\Delta x} = \lim_{\Delta x \to 0^-} \frac{2\Delta x - \Delta x}{\Delta x} = 1.$$

If $\Delta x \to 0^+$ then

$$\lim_{\Delta x \to 0+} \frac{2\left(\frac{2}{3}\Delta x - \frac{1}{3}\Delta x\right) + \left|\frac{2}{3}\Delta x - \frac{1}{3}\Delta x\right|}{\Delta x} = \lim_{\Delta x \to 0+} \frac{\frac{2}{3}\Delta x + \frac{1}{3}\Delta x}{\Delta x} = 1.$$

Therefore $F'(0) = 1$. (Another way is to show that $F(x) = x$).

How has the counter-example been constructed? How might others be found?

Counter-Example 4.4.21

If a function $y = f(x)$ is defined on $[a, b]$, differentiable on (a, b) and $f(a) = f(b)$, then there exists a point $c \in (a, b)$ such that $f'(c) = 0$.

The function shown below is defined on $[0, 3]$, differentiable on $(0, 3)$ and $f(0) = f(3) = 2$ but there is no point c in $(0, 3)$ such that $f'(c) = 0$.

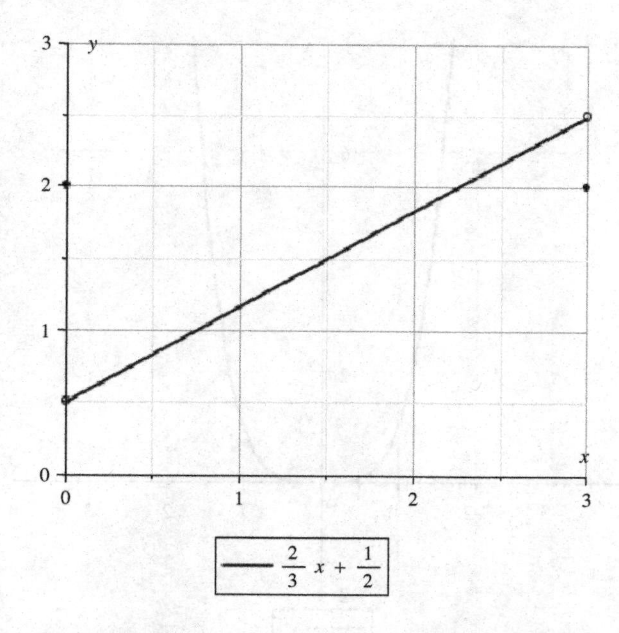

What if f was differentiable (from the appropriate side) at one of the end points: Would there still be a counter-example?

Counter-Example 4.4.22

If a function is twice-differentiable in a certain neighbourhood around $x = a$ and its second derivative is zero at that point then the point $(a,\ f(a))$ is a point of inflection for the graph of the function.

The function $y = x^4$ is twice differentiable on R and its second derivative is zero at the point $x = 0$ but the point (0, 0) is not a point of inflection.

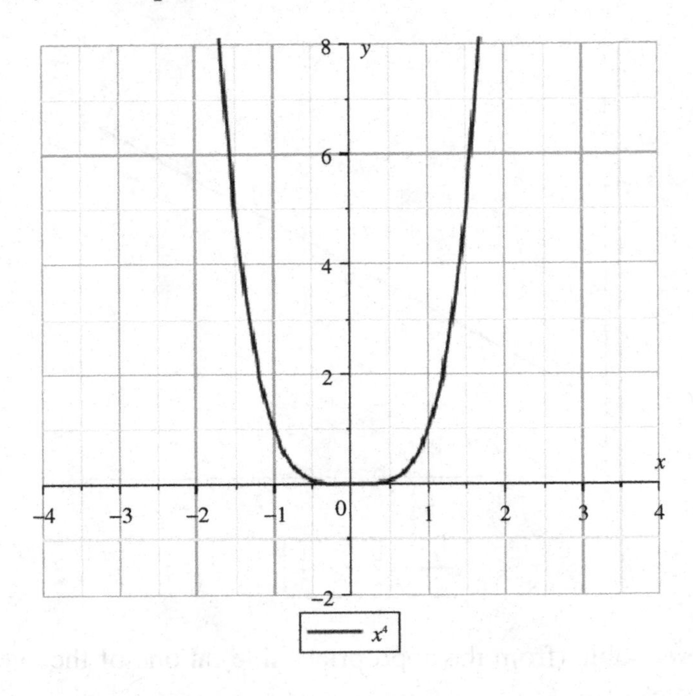

What conditions are necessary to be sure of an inflection point? What other functions share the same property as the example given? What is the most general function you can construct with this property?

Counter-Example 4.4.23

If a function $y = f(x)$ is differentiable at the point $x = a$ and the point $(a, f(a))$ is a point of inflection on the function's graph then the second derivative is zero at that point.

The function $y = x \times |x|$ is differentiable at $x = 0$ and the point (0, 0) is a point of inflection but the second derivative does not exist at $x = 0$. Although it looks a bit like $y = x^3$, when the two are juxtaposed, the difference is evident.

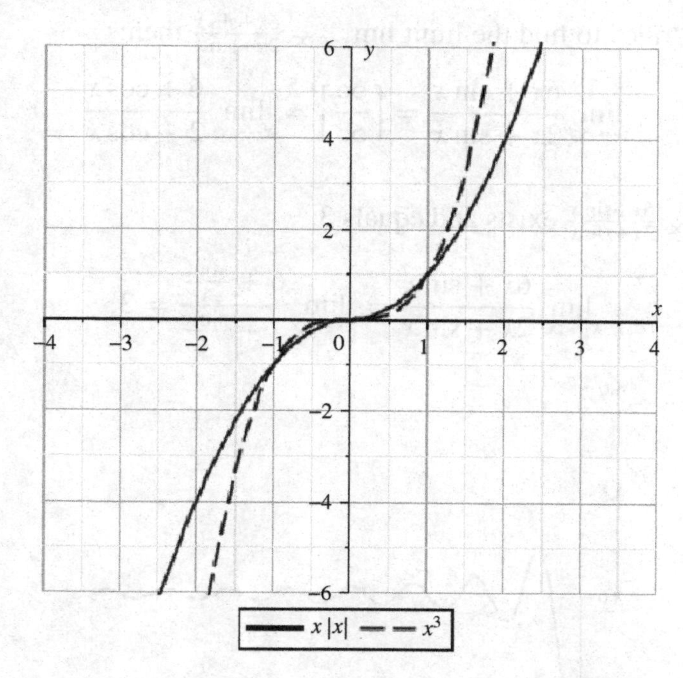

What is it about the example which allows it to be a counter-example? What general class of functions could be used in place of this particular example as a counter-example?

Counter-Example 4.4.24

If both functions $y = f(x)$ and $y = g(x)$ are differentiable on R then to evaluate the limit $\lim_{x\to\infty} \frac{f(x)}{g(x)}$ in the indeterminate form of type $\left[\frac{\infty}{\infty}\right]$ we can use the following rule:

$$\lim_{x\to\infty} \frac{f(x)}{g(x)} = \lim_{x\to\infty} \frac{f'(x)}{g'(x)}.$$

If we use the above "rule" to find the limit $\lim_{x\to\infty} \frac{6x+\sin x}{2x+\sin x}$ then:

$$\lim_{x\to\infty} \frac{6x+\sin x}{2x+\sin x} = \left[\frac{\infty}{\infty}\right] = \lim_{x\to\infty} \frac{6+\cos x}{2+\cos x}$$

is undefined.

But the limit $\lim_{x\to\infty} \frac{6x+\sin x}{2x+\sin x}$ exists and equals 3:

$$\lim_{x\to\infty} \frac{6x+\sin x}{2x+\sin x} = \lim_{x\to\infty} \frac{6+\frac{\sin x}{x}}{2+\frac{\sin x}{x}} = 3.$$

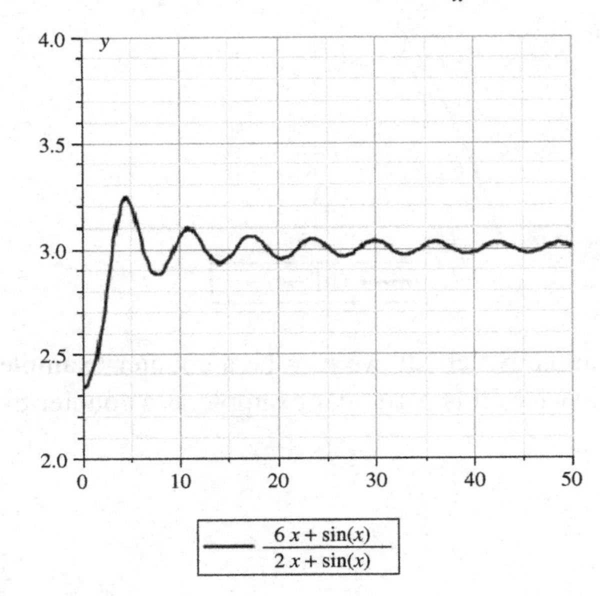

Comments. To make the above "rule" correct we need to add "if the limit $\lim_{x\to\infty} \frac{f'(x)}{g'(x)}$ exists or equals $\pm\infty$". This is the well-known l'Hospital's Rule for limits.

What is it about x and $\sin x$ which makes the example work as a counter-example? Can the example be modified to produce a counter-example to the corresponding conjecture when the limit is to some point c which is finite?

Counter-Example 4.4.25

If a function $y = f(x)$ is differentiable on (a, b) and $\lim_{x\to a^+} f'(x) = \infty$ then $\lim_{x\to a^+} f(x) = \infty$.

The function $y = \sqrt[3]{x}$ is differentiable on $(0, 1)$ and $\lim_{x\to 0^+} y'(x) = \lim_{x\to 0^+} \frac{1}{3\sqrt[3]{x^2}} = \infty$ but $\lim_{x\to 0^+} y(x) = \lim_{x\to 0^+} \sqrt[3]{x} = 0$.

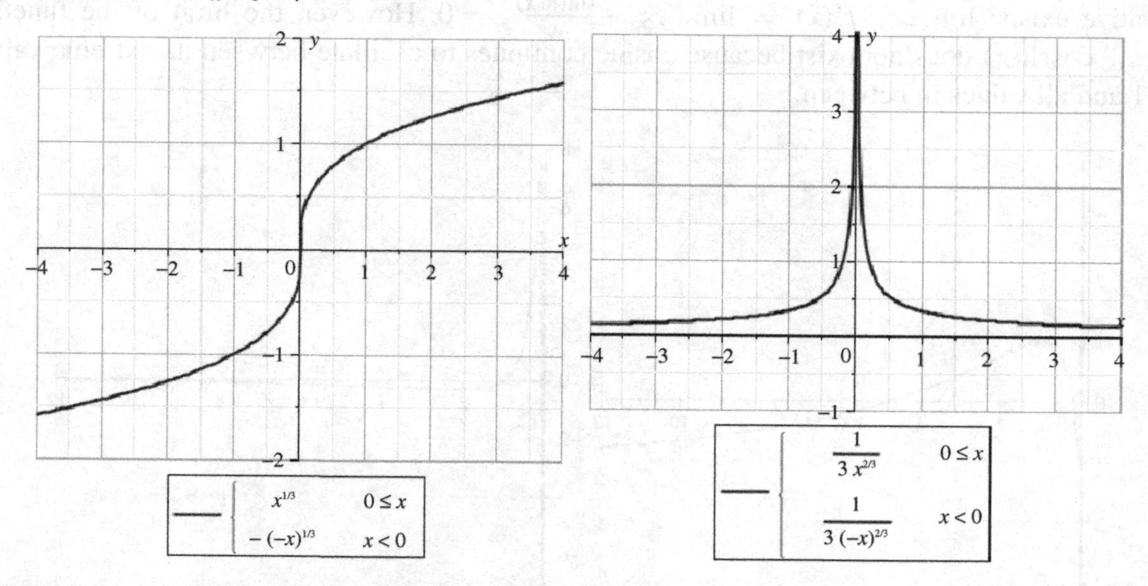

What other functions would serve as counter-examples?

Counter-Example 4.4.26

If a function $y = f(x)$ is differentiable on $(0, \infty)$ and $\lim_{x\to\infty} f(x)$ exists then $\lim_{x\to\infty} f'(x)$ also exists.

The function $f(x) = \frac{\sin(x^2)}{x}$ is differentiable on $(0, \infty)$ and $\lim_{x\to\infty} \frac{\sin(x^2)}{x} = 0$ but $\lim_{x\to\infty} f'(x) = \lim_{x\to\infty} \frac{2x^2\cos(x^2) - \sin(x^2)}{x^2}$ does not exist.

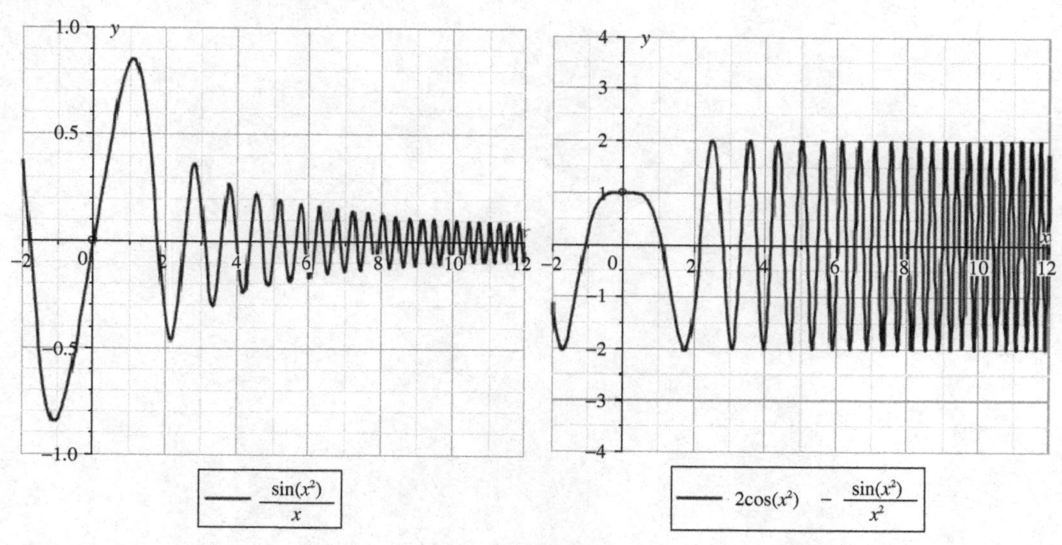

What role is played by the x^2 in this example? What else could it be? What does the counter-example say about slopes of functions tending to 0? Could the value 0 be altered to something else and still produce a counter-example?

Counter-Example 4.4.27

If a function $y = f(x)$ is differentiable and bounded on $(0, \infty)$ and $\lim_{x\to\infty} f'(x)$ exists then $\lim_{x\to\infty} f(x)$ also exists.

The function $f(x) = \cos(\ln x)$ is differentiable and bounded on $(0.\infty)$ and the limit of its derivative exists: $\lim_{x\to\infty} f'(x) = \lim_{x\to\infty} -\frac{\sin(\ln x)}{x} = 0$. However, the limit of the function $\lim_{x\to\infty} \cos(\ln x)$ does not exist because cosine continues to oscillate between its extreme values of ± 1 and all values in between.

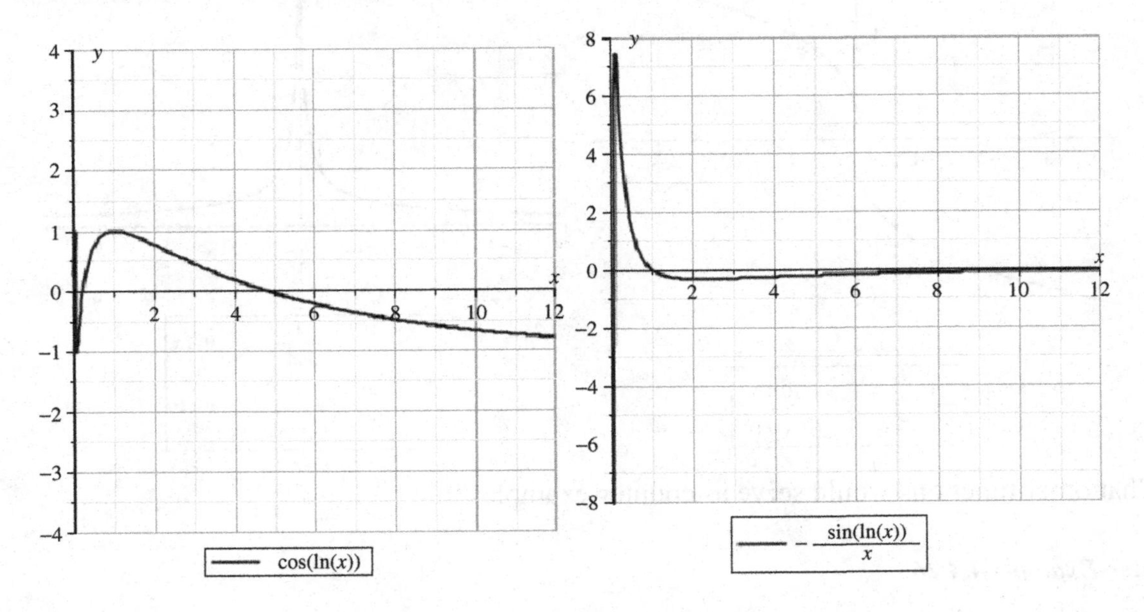

What role is played by $\ln x$? what functions could replace it and still be counter-examples? What needs to be changed to make the conjecture true?

Counter-Example 4.4.28

If a function $y = f(x)$ is differentiable at the point $x = a$ then its derivative is continuous at $x = a$.

The function

$$f(x) = \begin{cases} x^2 \sin \frac{1}{x}, & \text{if } x \neq 0 \\ 0, & \text{if } x = 0 \end{cases}$$

is differentiable at $x = 0$ but its derivative

$$f'(x) = \begin{cases} 2x \sin \frac{1}{x} - \cos \frac{1}{x}, & \text{if } x \neq 0 \\ 0, & \text{if } x = 0 \end{cases}$$

is discontinuous at $x = 0$.

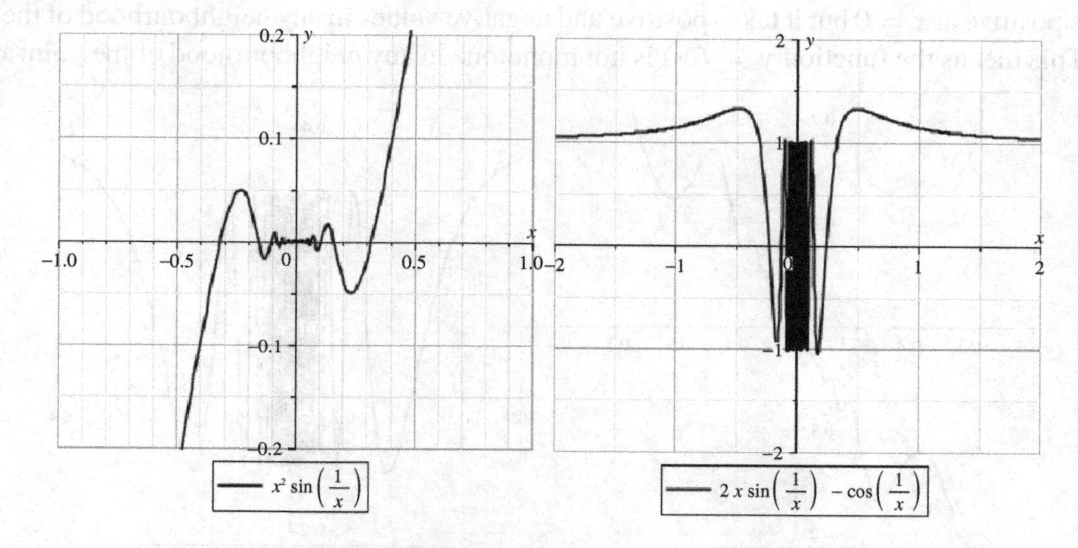

Could a counter-example be found which did not depend on rapid oscillation for its discontinuity at a point? Can the example be modified to produce counter-examples at finitely many points, and at infinitely many points?

Counter-Example 4.4.29

If the derivative of a function $y = f(x)$ is positive at the point $x = a$ then there is a neighbourhood about $x = a$ (no matter how small) where the function is increasing.

The function

$$f(x) = \begin{cases} x + 2x^2 \sin \dfrac{1}{x}, & \text{if } x \neq 0 \\ 0, & \text{if } x = 0 \end{cases}$$

has the derivative

$$f'(x) = \begin{cases} 1 + 4x \sin \dfrac{1}{x} - 2 \cos \dfrac{1}{x}, & \text{if } x \neq 0 \\ 1, & \text{if } x = 0 \end{cases}$$

which is positive at $x = 0$ but it takes positive and negative values in any neighbourhood of the point $x = 0$. This means the function $y = f(x)$ is not monotone in any neighbourhood of the point $x = 0$.

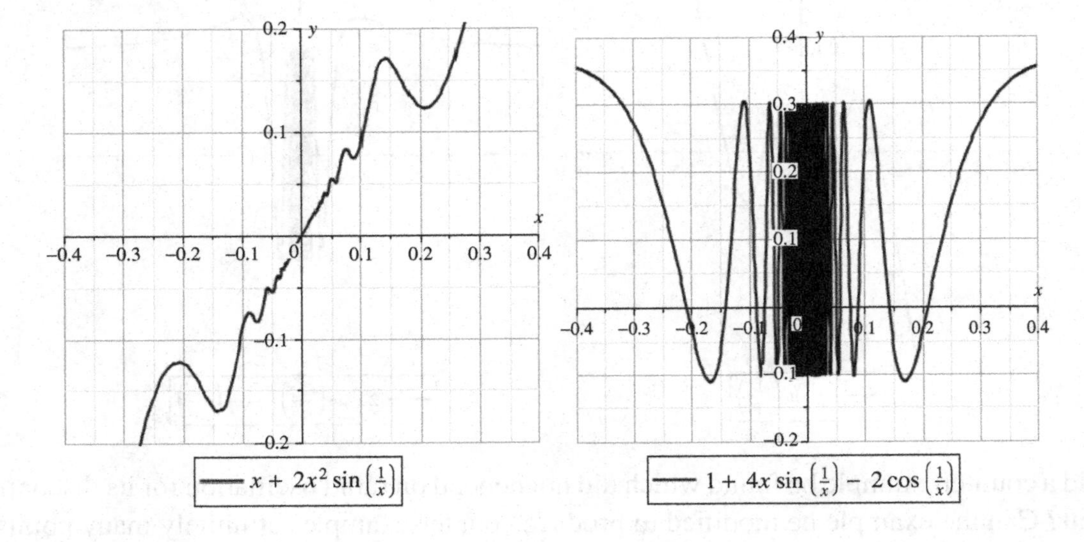

Counter-Example 4.4.30

If a function $y = f(x)$ is continuous on (a, b) and has a local maximum at the point $c \in (a, b)$ then in a sufficiently small neighbourhood of the point $x = c$ the function is increasing on the left and decreasing on the right from $x = c$.

The function

$$f(x) = \begin{cases} 2 - x^2\left(2 + \sin\dfrac{1}{x}\right), & \text{if } x \neq 0 \\ 2, & \text{if } x = 0 \end{cases}$$

is continuous on R. Since $x^2(2 + \sin\frac{1}{x})$ is positive for all $x \neq 0$ then $2 > 2 - x^2(2 + \sin\frac{1}{x})$. Therefore the function $y = f(x)$ has a local maximum at the point $x = 0$. But it is neither increasing on the left nor decreasing on the right in any neighbourhood of the point $x = 0$. To show this we can find the derivative $f'(x) = -4x - 2x\sin\frac{1}{x} + \cos\frac{1}{x}$; $x \neq 0$. The derivative takes both positive and negative values in any interval $(-\delta, 0) \cup (0, \delta)$ and therefore the function is not monotone in any interval $(-\delta, 0) \cup (0, \delta)$, where $\delta > 0$.

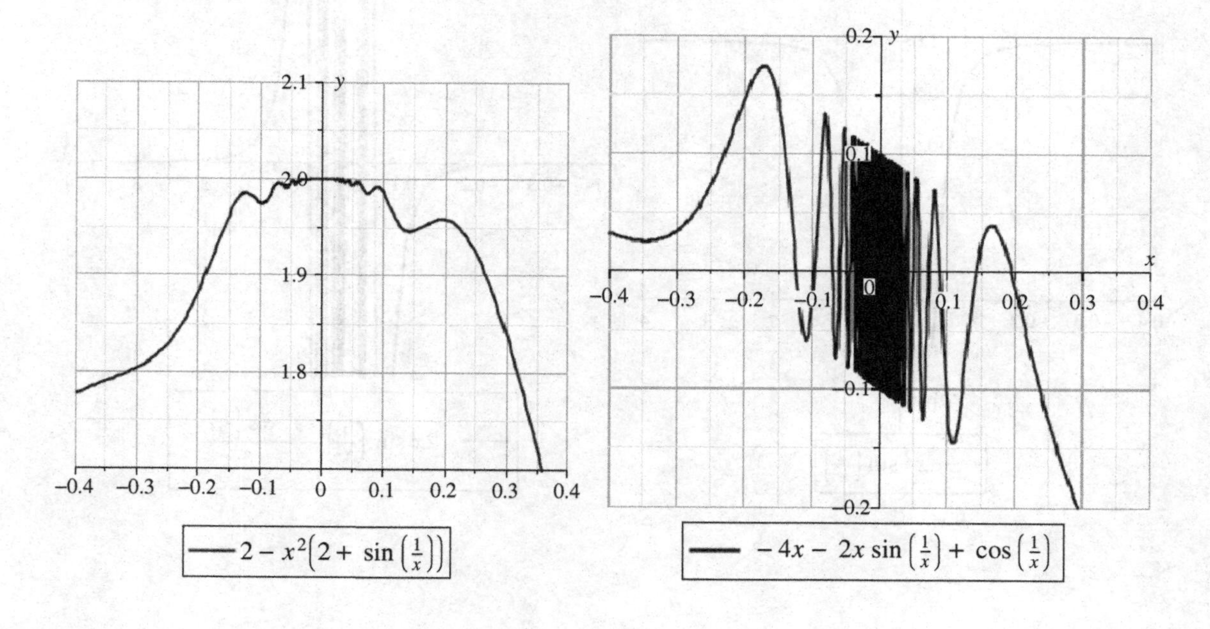

Counter-Example 4.4.31

If a function $y = f(x)$ is differentiable at the point $x = a$ then there is a certain neighbourhood of the point $x = a$ where the derivative of the function $y = f(x)$ is bounded.

The function

$$f(x) = \begin{cases} x^2 \sin \dfrac{1}{x^2}, & \text{if } x \neq 0 \\ 0, & \text{if } x = 0 \end{cases}$$

is differentiable at the point $x = 0$. Its derivative is

$$f'(x) = \begin{cases} 2x \sin \dfrac{1}{x^2} - \dfrac{2}{x} \cos \dfrac{1}{x^2}, & \text{if } x \neq 0 \\ 0, & \text{if } x = 0 \end{cases}.$$

The derivative of the function $y = f(x)$ is unbounded in any neighbourhood of the point $x = 0$.

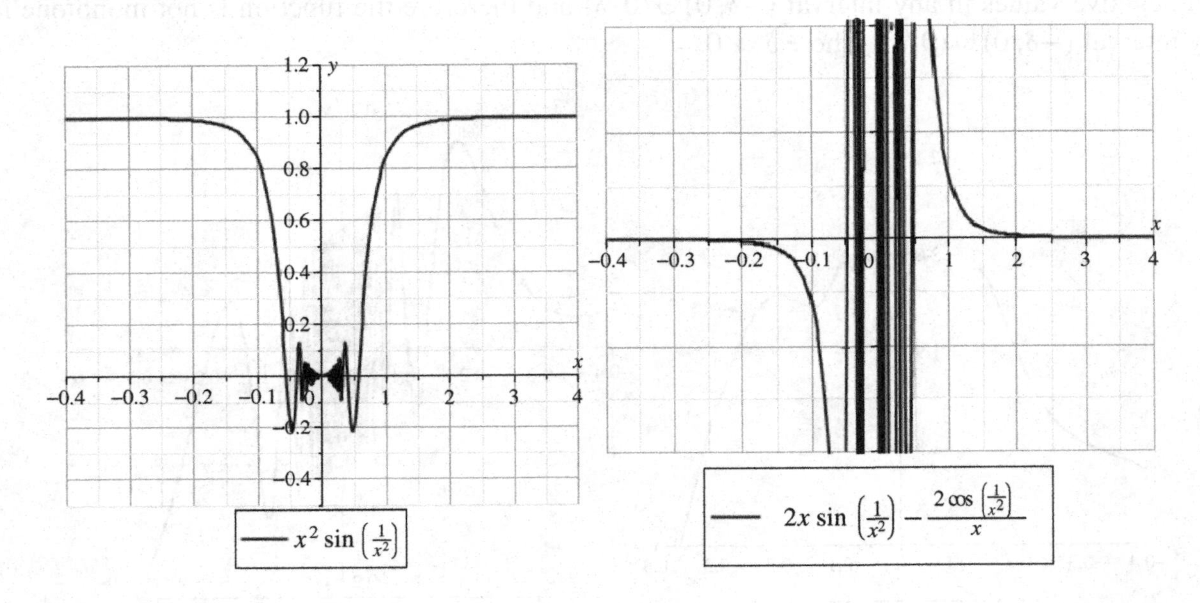

Counter-Example 4.4.32

If a function $y = f(x)$ at any neighbourhood of the point $x = a$ has points where $f'(x)$ does not exist then $f'(a)$ does not exist.

The function

$$f(x) = \begin{cases} x^2 \left|\cos \dfrac{\pi}{x}\right|, & \text{if } x \neq 0 \\ 0, & \text{if } x = 0 \end{cases}$$

in any neighbourhood of the point $x = 0$ has points where $f'(x)$ does not exist, however $f'(0) = 0$.

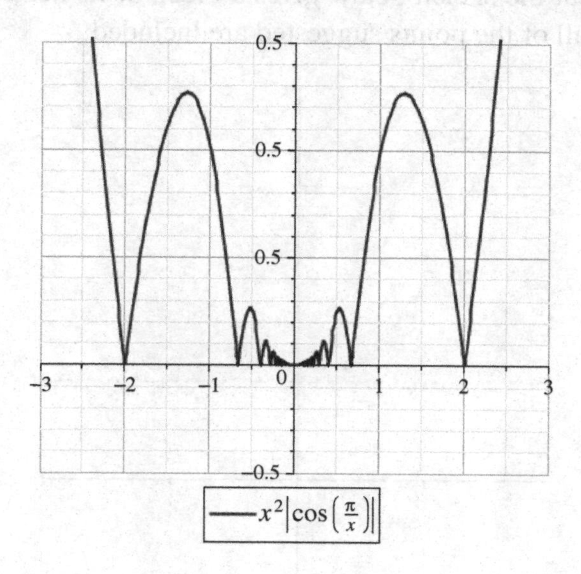

Counter-Example 4.4.33

A function cannot be differentiable only at one point in its domain and non-differentiable everywhere else in its domain.

The function

$$y = \begin{cases} 1 + x^2, & \text{if } x \text{ is rational} \\ 1, & \text{if } x \text{ is irrational} \end{cases}$$

is defined for all real x and differentiable only at the point $x = 0$. It is impossible to draw the graph of the function $y = f(x)$ but the sketch below gives an idea of its behaviour, where the fine dots indicate that some but not all of the points suggested are included.

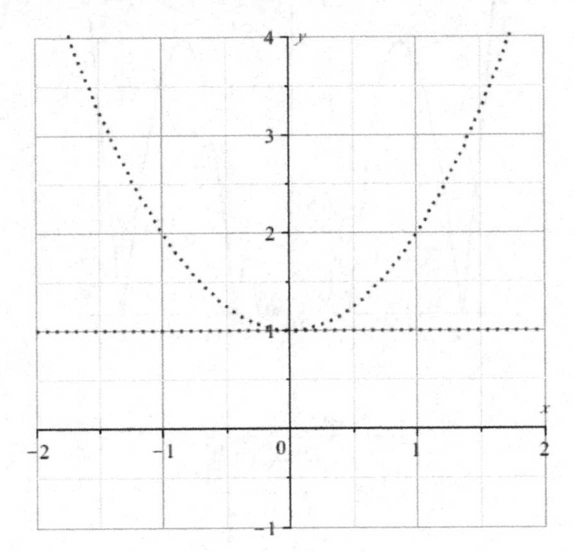

Counter-Example 4.4.34

A continuous function cannot be non-differentiable at every point in its domain.

The Weierstrass function can be defined as:

$$f(x) = \sum_{n=0}^{\infty} \left(\frac{1}{2}\right)^n \cos(3^n x).$$

If we take the first 3 and the first 7 terms in the sum we can begin to visualise the function:

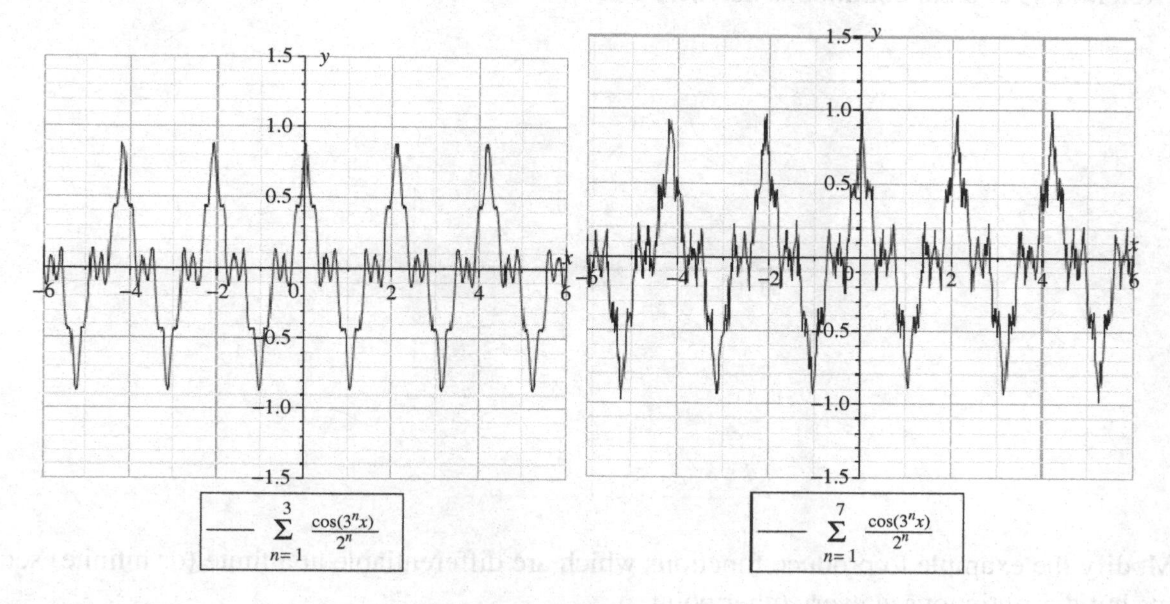

Comments. The Weierstrass function is the first known fractal. Another good example of a continuous curve that has a sharp corner at every point is the Koch's snowflake. We start with an equilateral triangle and build the line segments on each side according to a simple rule and repeat this process infinitely many times. The resulting curve is called Koch's curve and it forms the so-called Koch's snowflake. The first four iterations are shown below:

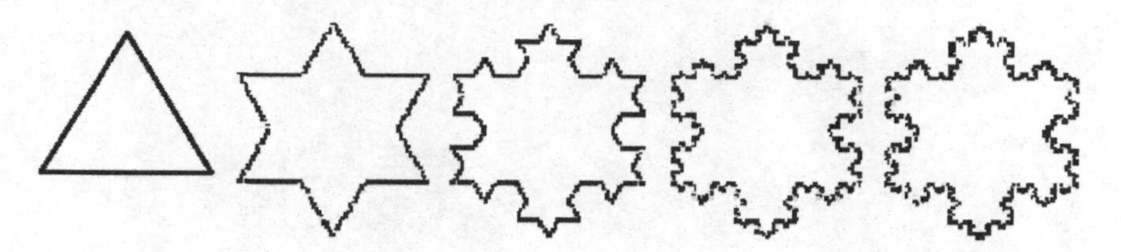

Counter-Example 4.4.35

A function cannot be differentiable at just one point without being at least continuous in a certain neighbourhood of that point.

Based on the function in 4.3.16,

$$f(x) = \begin{cases} x^2 & \text{if } x \text{ is rational} \\ -x^2 & \text{if } x \text{ is irrational} \end{cases}$$

is differentiable at 0 but continuous nowhere else.

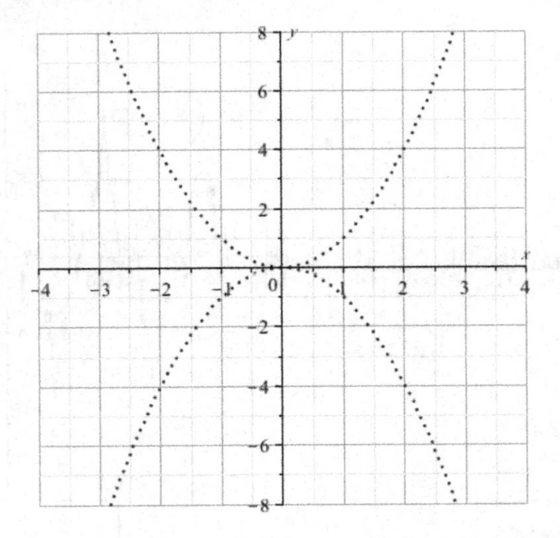

Modify the example to produce functions which are differentiable at a finite (or infinite) set of points but discontinuous at every other point.

4.5. Integral Calculus

We assume that by now readers are ready and eager to prompt themselves to develop, extend and vary the suggested counter-examples by asking questions similar to the questions from Sections 4.1–4.4. For example: What is it about the counter-example offered which makes it work? What alterations could you make to the counter-example and still produce a counter-example? How could the statement be modified to make it correct? Can you use another type of function as a counter-example? Can you construct the most general class of counter-examples? And others...

Counter-Example 4.5.1

If the function $y = F(x)$ is an antiderivative of a function $y = f(x)$ then $\int_a^b f(x)dx = F(b) - F(a)$.

The function $F(x) = \ln|x|$ is an antiderivative of the function $f(x) = \frac{1}{x}$ but the (improper) integral $\int_{-1}^{1} \frac{1}{x}\,dx$ does not exist.

Comments. To make the statement true we need to add that the function $y = f(x)$ must be continuous on $[a, b]$.

Counter-Example 4.5.2

If a function $y = f(x)$ is continuous on $[a, b]$ then the area enclosed by the graph of $y = f(x)$, OX, $x = a$ and $x = b$ numerically equals $\int_a^b f(x)dx$.

For any continuous function $y = f(x)$ that takes only negative values on $[a, b]$ the integral $\int_a^b f(x)dx$ is negative, therefore the area enclosed by the graph of $f(x)$, OX, $x = a$ and $x = b$ is numerically equal to $-\int_a^b f(x)dx$, or $|\int_a^b f(x)dx|$.

Counter-Example 4.5.3

If $\int_a^b f(x)dx \geq 0$ then $f(x) \geq 0$ for all $x \in [a, b]$.

$\int_{-1}^{2} xdx = \frac{3}{2} > 0$ but the function $y = x$ takes both positive and negative values on $[-1, 2]$.

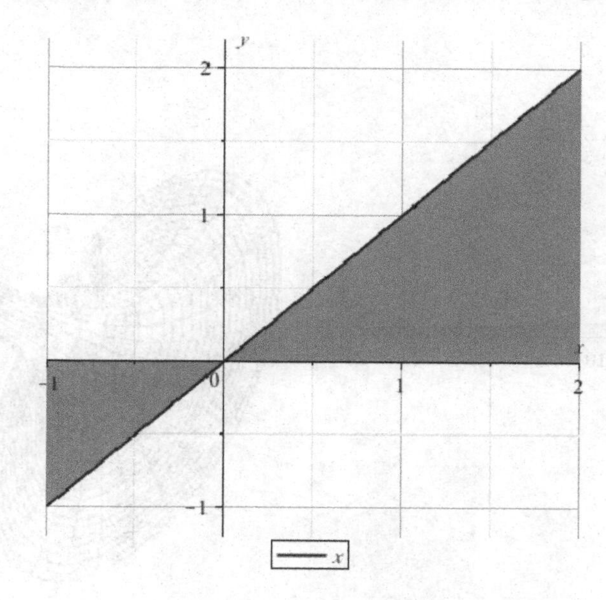

Counter-Example 4.5.4

If $y = f(x)$ is a continuous function and k is any constant then:

$$\int kf(x)dx = k\int f(x)dx.$$

If $k = 0$ then the left-hand side is:

$$\int 0f(x)dx = \int 0dx = C,$$

where C is an arbitrary constant. The right-hand side is:

$$0\int f(x)dx = 0.$$

This suggests that C is always equal to zero, but this contradicts the nature of an arbitrary constant.

Comments. The property is valid only for non-zero values of the constant k.

Counter-Example 4.5.5

A plane figure of infinite area rotated around an axis always produces a solid of revolution of infinite volume.

The figure enclosed by the graph of the function $y = \frac{1}{x}$, the x-axis and the straight line $x = 1$ is rotated about the x-axis.

The area is infinite:

$$\int_1^{\infty} \frac{1}{x}dx = \lim_{b\to\infty} (\ln b - \ln 1) = \infty$$

(square units), but the volume is finite:

$$\pi\int_1^{\infty} \frac{1}{x^2}dx = -\pi \lim_{b\to\infty} \left(\frac{1}{b} - 1\right) = \pi \text{ (cubic units).}$$

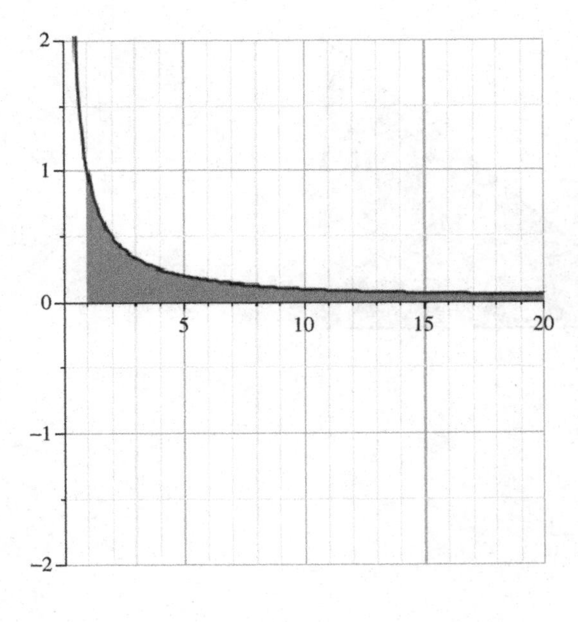

Counter-Example 4.5.6

If a function $y = f(x)$ is defined for any $x \in [a, b]$ and $\int_a^b |f(x)|dx$ exists then $\int_a^b f(x)dx$ exists.

The function

$$f(x) = \begin{cases} 1, & \text{if } x \text{ is rational} \\ -1, & \text{if } x \text{ is irrational} \end{cases}$$

is defined for any real x.

$|f(x)| = 1$ and therefore $\int_a^b |f(x)|dx = b - a$ but $\int_a^b f(x)dx$ does not exist. Let us show this using the definition of the definite integral.

Let $[a, b]$ be any closed interval. We divide the interval into n subintervals and find the limit of the integral sums:

$$S = \lim_{\max \Delta x_i \to 0} \sum_{i=0}^{n-1} f(c_i)\Delta x_i.$$

If on any subinterval we choose c_i equal to a rational number then $S = b - a$. If on any subinterval we choose c_i equal to an irrational number then $S = a - b$. So, the limit of the integral sums depends on the way we choose c_i and for this reason the definite integral of $f(x)$ on $[a, b]$ does not exist.

Counter-Example 4.5.7

If neither of the integrals $\int_a^b f(x)dx$ and $\int_a^b g(x)dx$ exist then the integral $\int_a^b (f(x) + g(x))dx$ does not exist.

For the functions

$$f(x) = \begin{cases} 1, & \text{if } x \text{ is rational} \\ -1, & \text{if } x \text{ is irrational} \end{cases} \quad \text{and} \quad g(x) = \begin{cases} -1, & \text{if } x \text{ is rational} \\ 1, & \text{if } x \text{ is irrational} \end{cases}$$

the integrals $\int_a^b f(x)dx$ and $\int_a^b g(x)dx$ do not exist (see the previous example 4.5.6) but the integral $\int_a^b (f(x) + g(x))dx$ exists and equals 0.

Counter-Example 4.5.8

If $\lim_{x\to\infty} f(x) = 0$ then $\int_a^\infty f(x)dx$ converges.

The limit $\lim_{x\to\infty} \frac{1}{x} = 0$ but the integral $\int_1^\infty \frac{1}{x}dx$ diverges. Note that the divergence cannot be "seen" in the graph.

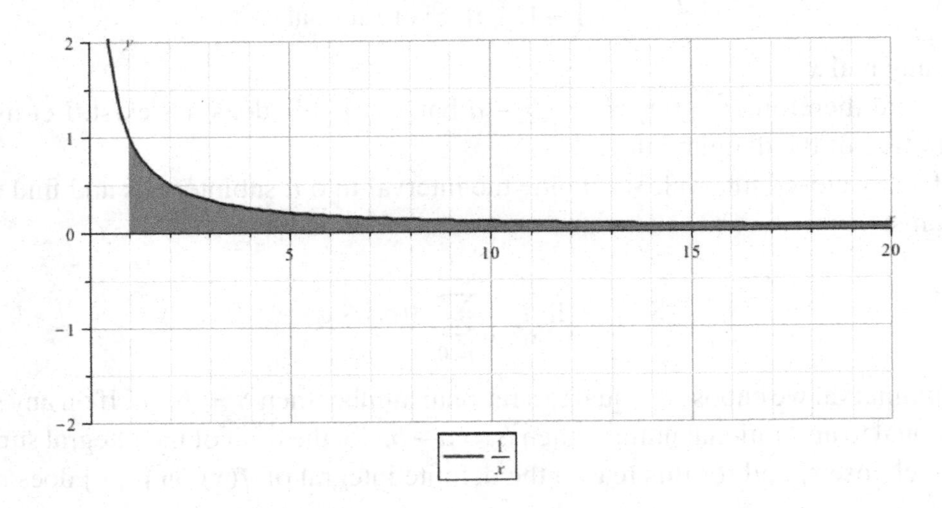

Counter-Example 4.5.9

If the integral $\int_a^\infty f(x)dx$ diverges then the function $y = f(x)$ is not bounded.

The integral of a non-zero constant $\int_a^\infty k\,dx$ is divergent but the function $y = k$ is bounded.

Counter-Example 4.5.10

If a function $y = f(x)$ is continuous and non-negative for all real x and $\sum_{n=1}^\infty f(n)$ is finite then $\int_1^\infty f(x)dx$ converges.

The function $y = |\sin \pi x|$ is continuous and non-negative for all real x and $\sum_{n=1}^\infty |\sin \pi n| = 0$ but $\int_1^\infty |\sin \pi x|dx$ diverges. Note that the divergence can be "seen" because of a constant area repeated infinitely often.

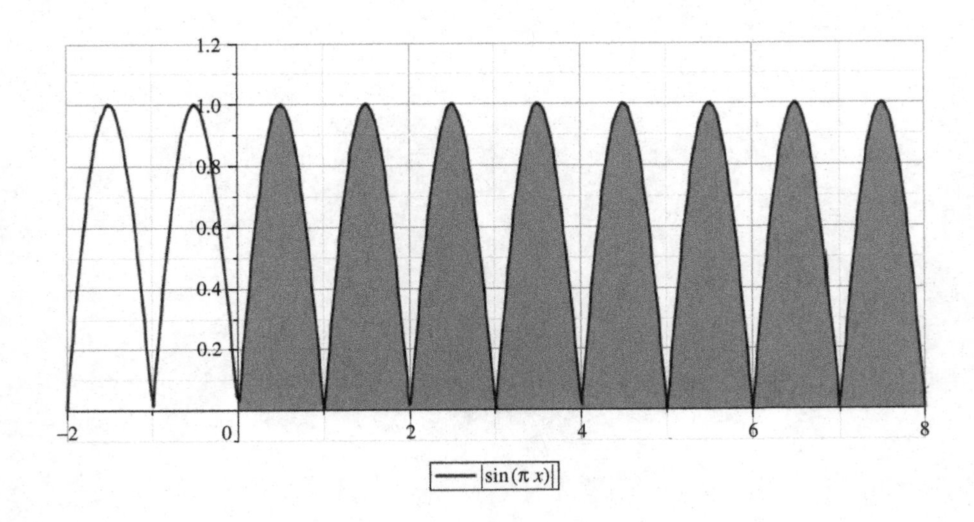

Counter-Example 4.5.11

If both integrals $\int_a^\infty f(x)dx$ and $\int_a^\infty g(x)dx$ diverge then the integral $\int_a^\infty (f(x)+g(x))dx$ also diverges.

Both integrals $\int_1^\infty \frac{1}{x}dx$ and $\int_1^\infty \frac{1-x}{x^2}dx$ diverge but the integral $\int_1^\infty (\frac{1}{x}+\frac{1-x}{x^2})dx = \int_1^\infty \frac{1}{x^2}dx$ converges.

Counter-Example 4.5.12

If a function $y = f(x)$ is continuous and $\int_a^\infty f(x)dx$ converges then $\lim_{x\to\infty} f(x) = 0$.

The Fresnel integral $\int_0^\infty \sin x^2 dx$ converges but $\lim_{x\to\infty} \sin x^2$ does not exist. Note that the convergence cannot be "seen" in the graph, and regions below the axis contribute a negative amount to the integral.

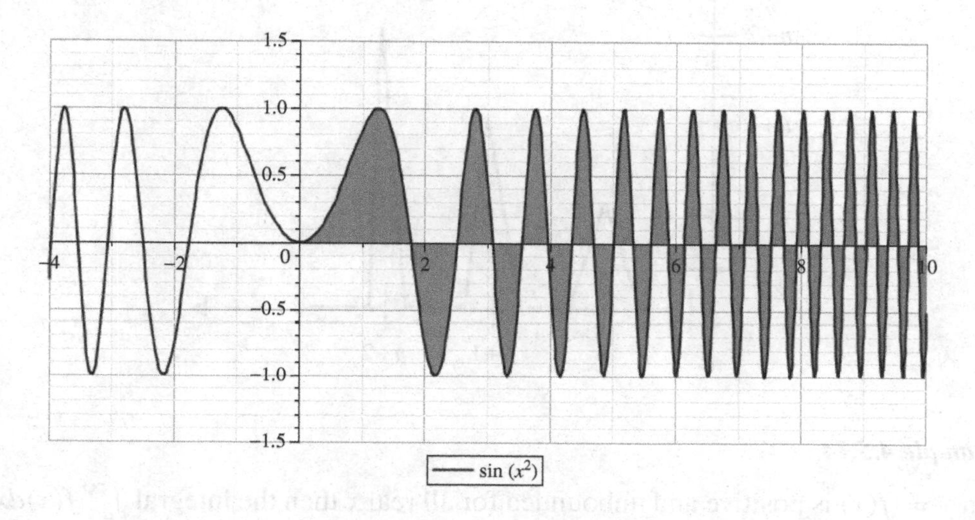

Counter-Example 4.5.13

If a function $y = f(x)$ is continuous and non-negative and $\int_a^\infty f(x)dx$ converges then $\lim_{x\to\infty} f(x) = 0$.

We will use the idea of area. Over every natural n we can construct triangles of area $\frac{1}{n^2}$ so that the total area equals $\sum_{n=a}^{\infty} \frac{1}{n^2}$, which is a finite number. The height of each triangle is n and the base is $\frac{2}{n^3}$.

The integral $\int_a^\infty f(x)dx$ converges since it is numerically equal to the total area $\sum_{n=a}^{\infty} \frac{1}{n^2}$. As one can see from the graph below the function (in bold) is continuous and non-negative but $\lim_{x\to\infty} f(x)$ does not exist.

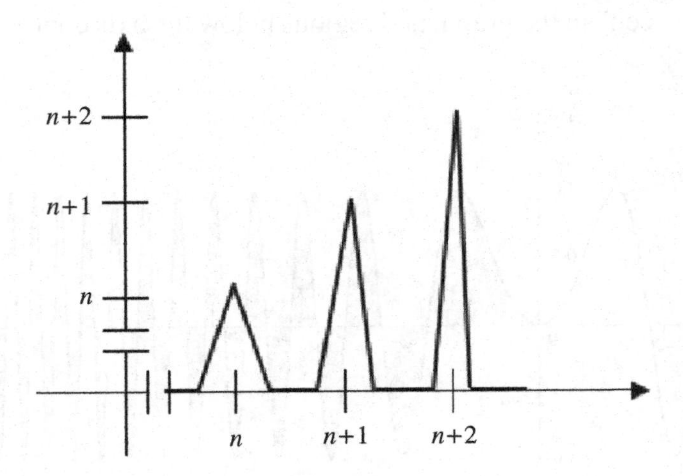

Counter-Example 4.5.14

If a function $y = f(x)$ is positive and unbounded for all real x then the integral $\int_a^\infty f(x)dx$ diverges.

We will use the idea of area. Over every natural n we can construct a rectangle with the height n and the base $\frac{1}{n^3}$ so the area is $\frac{1}{n^2}$. Make the function 0 elsewhere. Then the total area equals $\sum_{n=a}^{\infty} \frac{1}{n^2}$, which is a finite number. The positive and non-bounded function equals n on the interval of length $\frac{1}{n^3}$ around $x = n$, where n is a natural number. Since the integral $\int_a^\infty f(x)dx$ numerically equals the total area $\sum_{n=a}^{\infty} \frac{1}{n^2}$ it converges.

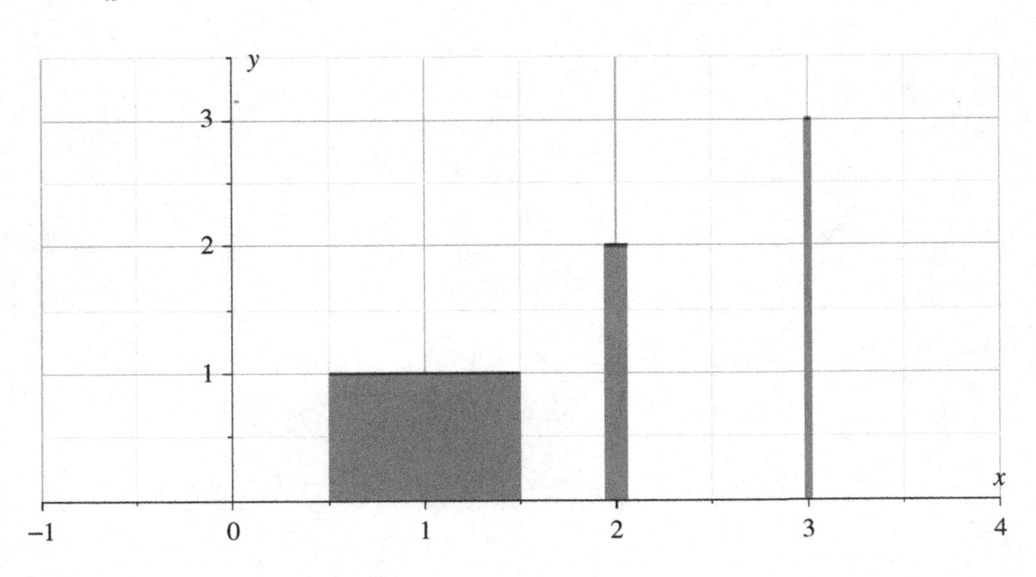

Counter-Example 4.5.15

If a function $y = f(x)$ is continuous and not bounded for all real x then the integral $\int_0^\infty f(x)dx$ diverges.

The function $y = x \sin x^4$ is continuous and unbounded for all real x, but the integral $\int_0^\infty x \sin x^4 dx$ converges (making the substitution $t = x^2$ yields the Fresnel integral $\frac{1}{2}\int_0^\infty \sin t^2 dt$ which is convergent). Note that the convergence cannot be "seen" in the graph.

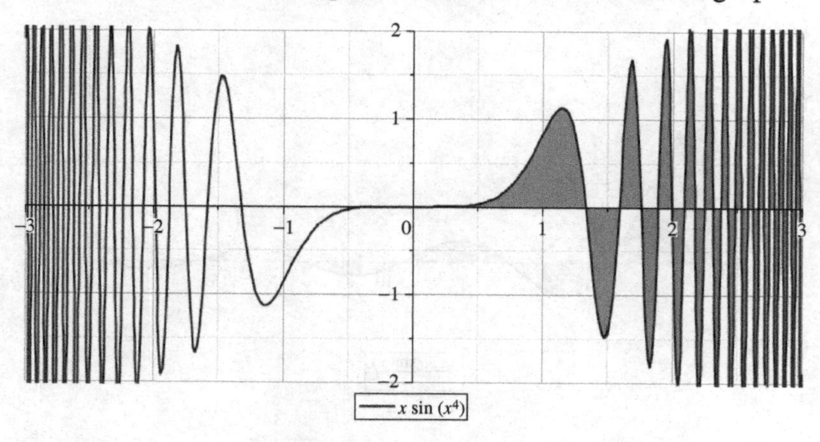

Counter-Example 4.5.16

If a function $y = f(x)$ is continuous on $[1, \infty)$ and $\int_1^\infty f(x)dx$ converges then $\int_1^\infty |f(x)|dx$ also converges.

The function $y = \frac{\sin x}{x}$ is continuous on $[1, \infty)$ and $\int_1^\infty \frac{\sin x}{x} dx$ converges but $\int_1^\infty |\frac{\sin x}{x}|dx$ diverges. Note that the convergence and divergence cannot be "seen" in the graphs.

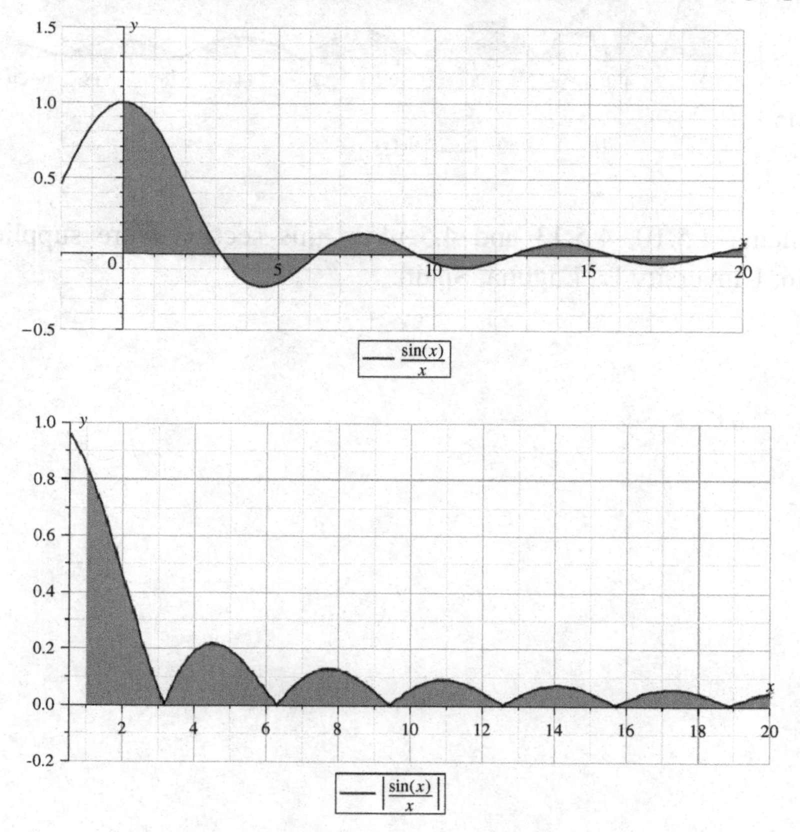

Counter-Example 4.5.17

If the integral $\int_a^\infty f(x)dx$ converges and a function $y = g(x)$ is bounded then the integral $\int_a^\infty f(x)g(x)dx$ converges.

The integral $\int_0^\infty \frac{\sin x}{x}dx$ converges and the function $g(x) = \sin x$ is bounded but the integral $\int_0^\infty \frac{\sin^2 x}{x}dx$ diverges. Note that the convergence and divergence cannot be "seen" in the graph.

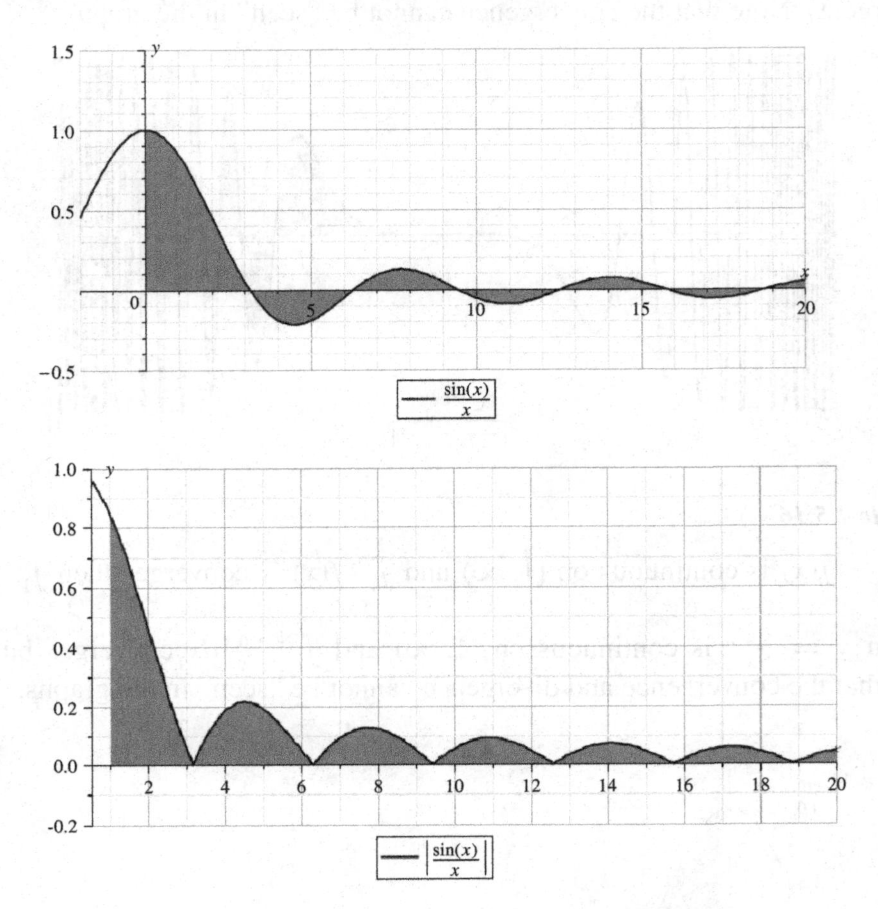

Comments. Statements 4.5.10, 4.5.13 and 4.5.14 in this section were supplied by Alejandro S. Gonzalez-Martin, University La Laguna, Spain.

BIBLIOGRAPHY

Bermudez, C. G. (2004). Counterexamples in calculus teaching. Paper presented at the 10th International Congress on Mathematics Education (ICME-10) (Copenhagen, Denmark).

Dahlberg, R. and Housman, D. (1997). Facilitating learning events through example generation. *Educational Studies in Mathematics*, **33**, pp. 283–299.

Feynman, R. (1985). *"Surely you're joking, Mr Feynman!": Adventures of a Curious Character* (W. W. Norton: New York).

Gelbaum, B. R. and Olmstead, J. M. H. (1964). *Counterexamples in Analysis* (Holden-Day: San Francisco).

Gruenwald, N. and Klymchuk, S. (2003). Using counter-examples in teaching Calculus. *The New Zealand Mathematics Magazine*, **40**(2), pp. 33–41.

Halmos, P. (1983). D. Sarasen and L. Gillman (Eds.) *Selecta: Expository Writing* (Springer-Verlag: New York).

Halmos, P. (1975). The problem of learning to teach, *American Mathematical Monthly*, **82**(5), pp. 466–76.

Halmos, P. (1985). *I Want to be a Mathematician: An Automathography* (Springer-Verlag, New York).

Halmos, P. (1994). What is teaching? *American Mathematical Monthly*, **101**(9), pp. 848–854.

Hauchecorne, B. (1988). Les contre-exemples en mathematiques (Ellipses, Paris).

Klymchuk, S. (2004). Counter-Examples in Calculus (Maths Press, New Zealand).

Klymchuk, S. (2005). Counter-examples in teaching/learning of Calculus: Students' performance. *The New Zealand Mathematics Magazine*, **42**(1), pp. 31–38.

Lakatos, I. (1976). *Proofs and Refutations: The Logic of Mathematical Discovery* (Cambridge University Press: Cambridge).

MacHale, D. (1980). The Predictability of Counterexamples, *American Mathematical Monthly*, **87**, pp. 752.

Marton, F. and Booth, S. (1997). *Learning and Awareness* (Lawrence Erlbaum: Mahwah).

Mason, J. (2003). Structure of Attention in the Learning of Mathematics, in J. Novotná (Ed.) *Proceedings, International Symposium on Elementary Mathematics Teaching* (Charles University, Prague), pp. 9–16.

Goldenberg, P. and Mason, J. (2008). Spreading Light on and with Example Spaces. *Educational Studies in Mathematics*, OnLineFirst.

Mason, J. and Watson, A. (2001). Getting students to create boundary examples. *MSOR Connections*, 1, **1**, pp. 9–11.

Michener, E. (1978). Understanding Understanding Mathematics. *Cognitive Science*, **2**, pp. 361–383.

Papert, S. (1993). *The Children's Machine: Rethinking School in the Age of the Computer* (Basic Books: New York).

Peled, I. and Zaslavsky, O. (1997). Counter-Examples that (only) Prove and Counter-Examples that (also) Explain. *FOCUS on Learning Problems in Mathematics*, **19**(3), pp. 49–61.

Pólya, G. (1962). *Mathematical Discovery: On Understanding, Learning, and Teaching Problem Solving* (Wiley: New York).

Runesson, U. (2005). Beyond Discourse and Interaction. Variation: A critical aspect for teaching and learning mathematics. *Cambridge Journal of Education*, **35**(1), pp. 69–88.

Selden, A. and Selden, J. (1998). The Role of Examples in Learning Mathematics. *Research Sampler*, MAAOnLine, February 20, 1998 (accessed December 2005).

Tall, D. (1991). The psychology of advanced mathematical thinking. In: Tall (ed.) *Advanced Mathematical Thinking* (Dordrecht: Kluwer), pp. 3–21.

Watson, A. and Mason, J. (2005). *Mathematics as a Constructive Activity: The Role of Learner-Generated Examples* (Erlbaum: Mahwah).

Wilson, P. (1986). Feature frequency and the use of negative instances in a geometric task, *Journal for Research in Mathematics Education*, **17**, pp. 130–139.

Whitehead, A. (1911). *An Introduction to Mathematics* (reprinted 1948) (Oxford University Press: Oxford).

Zaslavsky, O. and Ron, G. (1998). Students' understanding of the role of counter-examples, *Proceedings of the 22nd Conference of the International Group for the Psychology of Mathematics Education*. **1**, pp. 225–232 (Stellenbosch, South Africa).